TRELLIS CODING

Books of Related Interest from IEEE Press . . .

INTRODUCTION TO NONPARAMETRIC DETECTION WITH APPLICATIONS
An IEEE Press Classic Reissue
Jerry Gibson and James L. Melsa
1996   Hardcover   256 pp   ISBN 0-7803-1161-2   PC5634

REED-SOLOMON CODES AND THEIR APPLICATIONS
Stephen B. Wicker and Vijay K. Bhargava
1994   Hardcover   336 pp   ISBN 0-7803-1025-X   PC3749

MULTIPLE ACCESS COMMUNICATIONS: Foundations for Emerging Technologies
Edited by Norman Abramson
1993   Hardcover   528 pp   ISBN 0-87942-292-0   PC2873

TRELLIS CODING

Christian Schlegel
University of Utah
University of Texas

Contribution by Lance Perez
University of Nebraska

IEEE
PRESS

The Institute of Electrical and Electronics Engineers, Inc., New York

This book and other books may be purchased at a discount
from the publisher when ordered in bulk quantities. Contact:

IEEE Press Marketing
Attn: Special Sales
Piscataway, NJ 08855-1331
Fax: (908) 981-9334

For more information about IEEE PRESS products,
visit the IEEE Home Page: http://www.ieee.org/

Printed in the United States of America

10   9   8   7   6   5   4   3   2   1

ISBN 0-7803-1052-7
IEEE Order Number: PC4069

Library of Congress Cataloging-in-Publication Data

Schlegel, Christian (date)
    Trellis Coding / Christian Schlegel: with a chapter contributed
  by Lance C. Perez
        p.   cm.
    Includes bibliographical references and index.
    ISBN 0-7803-1052-7
    1. Error-correcting codes (Information theory)   I. Title
  TK5102.96.S35   1997
621.38'--dc20   96-43184
                                                          CIP

To my friends

Ajay, Alex, Are, Bernhard,
Buddy, Cheryl, Christina,
Christoph, Daniel, Dennis,
Dölf, Dominik, Ernst, Frank,
Gabi, Gail, Graham, Guido,
Joe, Joël, John, Jürg, Karin,
Ken, Lance, Lars, Lei, Marc,
Mark, Maria, Marlene,
Martin, Margrit, Melissa,
Mike, Nancy, Paul, Peter,
Phil, Rhonda, Richard, Rita,
Robert, Ruedi, Stephen,
Sumit, Tor, Uli, Weimin

CONTENTS

PREFACE

This book is about trellis coding, a branch of error control coding that proved remarkably successful in theory and practice alike. The book has grown out of my involvement with the subject over the years, starting with my graduate work at the University of Notre Dame. During that time in the latter 1980s, trellis coding, especially trellis-coded modulation (TCM), became a "hot" research topic, and most of the material presented in this book was compiled and published over a span of only a few years, attesting to the excitement in the research community during that time. One of the most far-reaching outgrowths of trellis coding, TCM, is an inventive marriage of convolutional coding with modulation. TCM made it possible to build communications systems with greatly superior data rates and reliability. In the last decade, the popularity of TCM has exploded while the telecommunications industry has seen one of the fastest migrations from the theorist's desk to commercial products and international standards. Trellis coding is a narrow specialty within the large field of telecommunications. Nonetheless, it has become a topic of significant importance and an essential component in the repertoire of the modern communications engineer.

As with most technical subjects, there are many ways of approaching trellis coding, and I have decided to start out by laying a rather general theoretical basis in Chapters 1 and 2. It is the goal of this book to provide a basic well-rooted understanding of trellis coding that brings together various viewpoints. Some sections are rather theoretical and others are more practice oriented, presenting algorithms and simulation results. The emphasis is on a theoretical, unified view, and I am aware that many interesting details of a more application-oriented nature had to be omitted. However, the book contains many simulation results that illustrate the theory and design of the codes and algorithms. Theoretical excursions are taken if they illuminate particular aspects of the theory or provide additional insight.

There are several ways of working through the book, and the study map on the next page illustrates some of my suggestions. The most straightforward route to gain a basic understanding follows along Chapters 1 and 2, then through the first part of Chapter 3 to Chapters 5 and 6. The first sections of Chapter 3 introduce TCM and its original designs. Chapter 5 discusses the error performance of trellis codes and its calculation, and Chapter 6 deals with the problem of decoding trellis codes, presenting the most important decoding algorithms. A more leisurely route leads through the second part of Chapter 3, introducing lattices, the lattice formulation of trellis codes, rotational invariance, and the important topic of geometric uniformity. From there the reader may proceed along the main route or go directly to Chapter 7. It connects the two large camps of error control coding, namely block codes and trellis codes, by showing that block codes have a code trellis too. The reader interested in convolutional codes may want to take the detour via Chapter 4, which explores in detail some mathematical properties of convolutional encoders and their minimal representations. The final topic of Turbo codes in Chapter 8 is best approached from the main study route, since its exposition relies on the concept of the distance spectrum from Chapter 5 and the maximum a posteriori (MAP) decoder from Chapter 6.

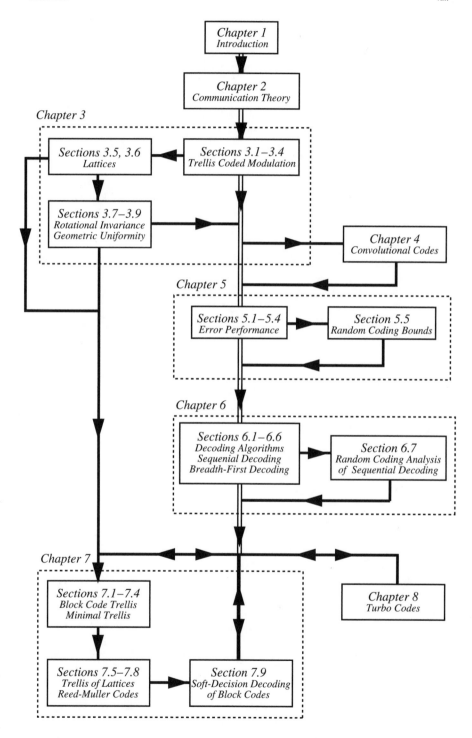

ACKNOWLEDGMENTS

I am indebted to a number of friends and colleagues who have helped me along this project with advice and support. I specifically wish to mention Lance Perez for agreeing to author Chapter 8 on turbo codes, Weimin Zhang for supplying or verifying many of the TCM simulation and code construction results, and Ajay Dholakia and Tor Aulin for carefully reviewing early versions of the manuscript and giving valuable suggestions and corrections along the way. I particularly appreciated their continual encouragement. Finally, I wish to thank Dan Costello for introducing me to this fascinating topic.

CHAPTER 1

INTRODUCTION

1.1 MODERN COMMUNICATIONS—MIGRATION TOWARD DIGITAL

With the advent of high-speed logic circuits and very large scale integration (VLSI), data processing and storage equipment has inexorably moved toward employing digital techniques. In digital systems data is encoded into strings of zeros and ones, corresponding to the on and off states of semiconductor switches. This is bringing about some fundamental changes in how information is processed. All real data is analog in one form or another, and this is the only way we can perceive it with our senses. Analog information is encoded into a digital representation, e.g., into a string of ones and zeros. This book will not deal with any such encoding technique; suffice it to say that conversions from analog to digital and back again are processes that have become ubiquitous. As a noteworthy example we might mention the digital encoding of speech.

Digital information is treated differently in communications than analog information. Signal estimation becomes signal detection; that is, a communications receiver need not look for an analog signal and make a "best" estimate, it only needs to make a decision between, say, a one or a zero. Digital signals are more reliable in a noisy communications environment. They can usually be detected perfectly, as long as the noise levels are not too high. This allows us to restore digital data and, through error-correcting techniques, even correct errors. On the other hand, digital data can be encoded in such a way as to introduce dependency among a large number of symbols, thus enabling the receiver to make a more accurate detection of the symbols. This is called *error control coding*.

1

The digitization of data is convenient for a number of additional reasons. The design of signal processing algorithms for digital data seems much easier than designing analog signal processing algorithms. The abundance of such digital algorithms, including error control and correction techniques, combined with their ease of implementation in VLSI circuits, has led to many successful applications of error control coding in practice.

The first use of error control coding was for deep-space probes where we are confronted with a low-power communications channel with virtually unlimited bandwidth. On these data links, convolutional codes (Chapter 4) are used with sequential and Viterbi decoding (Chapter 6). The next successful application of error control coding was to storage devices, most notably the compact disk player, which employs powerful Reed-Solomon codes [1], since the raw error probability from the optical readout device is too large for high-fidelity sound reproduction. Another hurdle taken was the successful application of error control to bandwidth-limited telephone channels, where trellis-coded modulation (Chapter 3) was used to produce impressive improvements. Nowadays coding is routinely applied to satellite communications [2, 3], teletext broadcasting, computer storage devices, logic circuits, semiconductor memory systems, magnetic recording systems, audio-video systems, and modern mobile communications systems like the pan-European digital telephony standard GSM [4] and IS 95 [5], a new American digital cellular standard using spread-spectrum techniques.

Figure 1.1 shows the basic configuration of a point-to-point digital communications link. The data to be transmitted over this link can come from some analog source, in which case it must first be converted into digital format (digitized), or a digital information source. If this data is a speech signal, for example, the digitizer is a speech codec [6]. Usually the digital data is source-encoded to remove unnecessary redundancy from the data; that is, the source data is compressed. This source encoding has the effect that the digital data which enters the encoder has statistics which resemble that of a random symbol source with maximum entropy; that is, all the different digital symbols occur with equal likelihood and are statistically independent. The channel encoder operates on this compressed data and introduces controlled redundancy for transmission over the channel. The modulator converts the discrete channel symbols into waveforms which are transmitted through the waveform channel. The demodulator reconverts the waveforms into a discrete sequence of received symbols, and the decoder reproduces an estimate of the compressed input data sequence, which is subsequently reconverted into the original signal or data sequence.

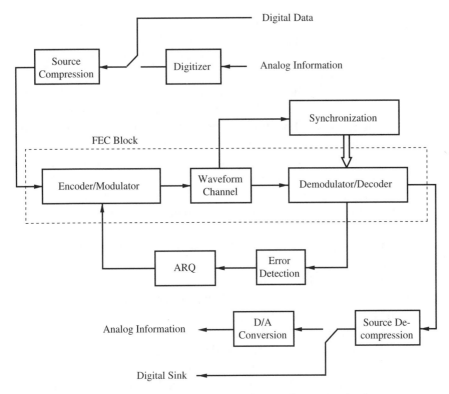

Figure 1.1 System diagram of a complete point-to-point communication system for digital data. The FEC block is the topic of this book.

An important ancillary function at the receiver is the synchronization process. We usually need to acquire carrier frequency and phase synchronization as well as symbol timing synchronization in order for the receiver to operate. Synchronization is not a topic of this book, and we will assume in most of our discussion that synchronization has been established (with the exception of phase synchronization in the case of rotationally invariant codes in Chapter 3). The book by Meyr and Ascheid [7] treats synchronization issues in detail. Since synchronization is a relatively slow estimation process, and data detection is a fast process, we usually have those two operations separated in real receiver implementations, as indicated in Figure 1.1.

Another important feature in some communication systems is *automatic repeat request* (ARQ). In ARQ the receiver also performs error detection and, through a return channel, requests retransmission of erroneous data blocks, or data blocks that cannot be reconstructed with sufficient confidence [8]. ARQ can usually improve the data transmission quality substantially, but the return channel needed for ARQ is not always available or may be impractical. For a deep-space probe on its way to the outer rim of our solar system, ARQ is not feasible since the return path takes too long (several hours). For speech-encoded signals ARQ is usually impossible for the same reason, since only a maximum speech delay is acceptable (about 200 ms typically, depending on the application). In broadcast systems, ARQ is ruled out for obvious reasons. Error control coding without ARQ is termed *forward error correction* or *control* (FEC). FEC is more difficult to perform than simple error detection and ARQ, but dispenses with the return channel. Often FEC and ARQ are both integrated into hybrid error control systems [8, 9] for data communications.

This book deals only with FEC, the dashed block in Figure 1.1. The reason we can do this relatively easily is due to the different functionalities of the various blocks just discussed. Each of them is mostly a separate entity with its own optimization strategy, and data is simply passed between the different blocks without much mutual interaction. A notable point in Figure 1.1 is that the encoder/modulator and the demodulator/decoder are combined operations. This basic novelty gave birth to trellis-coded modulation. This joint encoder/modulator design was first proposed by Wozencraft [10] and Massey [11], and then realized with stunning results by Ungerböck [12–14]. Even though trellis-coded modulation is not the only topic of this book, it is clearly a major motivation for its existence.

Since we will assume the encoder input data to be a sequence of independent, identically and uniformly distributed symbols (courtesy of the source compression), the single most important parameter to optimize for the FEC block is arguably the bit and/or symbol error rate, and we will adopt this as our criterion for the goodness of an FEC system. Note that this is not necessarily the most meaningful measure in all cases. Consider, for example, pulse-code-modulated (PCM) speech, where an error in the most significant bit is clearly more detrimental than an error in the least significant bit. Researchers have also looked at schemes with unequal error protection (e.g., [15]). However, those methods usually are a variation of the basic theme of obtaining a minimum error rate, and are not discussed in this book.

1.2   ERROR CONTROL CODING

The modern approach to error control in digital communications started with the ground-breaking work of Shannon [16], Hamming [17], and Golay

[18]. Shannon put down a theory that explained the fundamental limits on the efficiency of communications systems, and Hamming and Golay were the first to develop practical error control schemes. The new paradigm born was one in which errors are not synonymous with data that is irretrievably lost, but by clever design errors could be corrected or avoided altogether. This new thinking was revolutionary. But even though Shannon's theory promised that large improvements in the performance of communication systems could be achieved, practical improvements had to be excavated by intensive research over now nearly five decades. One reason for this lies in a curious shortcoming of Shannon's theory. It clearly states fundamental limits on communication efficiency, but its methodology provides no insight into how to actually achieve these limits. Coding theory, on the other hand, evolved from Hamming and Golay's work into a flourishing branch of applied mathematics [19].

Let us see where it all started. The most famous formula from Shannon's work is arguably the channel capacity of an ideal band-limited Gaussian channel,[1] which is given by

$$C = W \log_2 (1 + S/N) \quad \text{[bits/s]}. \tag{1.1}$$

In this formula C is the channel capacity, that is, the maximum number of bits that can be transmitted through this channel per unit of time (seconds [s]), W is the bandwidth of the channel, and S/N is the signal-to-noise power ratio at the receiver. Shannon's main theorem, which accompanies (1.1), asserts that error probabilities as small as desired can be achieved as long as the transmission rate R through the channel (in bits/second) is smaller than the channel capacity C. This can be achieved by using an appropriate encoding and decoding operation. However, Shannon's theory is silent about the structure of these encoders and decoders.

This new view was in marked contrast to early practices, which embodied the opinion that in order to reduce error probabilities the signal energy had to be increased; that is, S/N had to be improved. Figure 1.2 shows the error performance of quadrature phase-shift keying (QPSK), a popular modulation method for satellite channels (see Chapter 2) that allows data transmission up to 2 bits/symbol. The bit error probability of QPSK is shown as a function of S/N per dimension normalized per bit (see Section 1.3), henceforth called SNR. It is evident that an increased SNR provides a gradual decrease in error probability. This contrasts markedly with Shannon's theory, which promises zero(!) error probability at a spectral efficiency of 2 bits/s/Hz (Section 1.3), which is the maximum that

---

[1] The exact definitions of these basic communications concepts are given in Chapter 2.

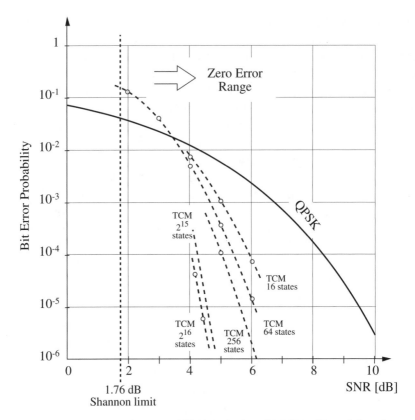

Figure 1.2 Bit error probability of QPSK and selected 8-PSK trellis-coded modulation (TCM) methods as a function of the normalized signal-to-noise ratio.

QPSK can achieve, as long as SNR > 1.5 (1.76 dB), shattering conventional wisdom prior to 1948.

Also shown in Figure 1.2 is the performance of several trellis-coded modulation (TCM) schemes using 8-ary phase-shift keying (8-PSK) (Chapter 3), and the improvement of coding becomes evident. The difference in SNR for an objective target bit error rate between a coded system and an uncoded system is termed the *coding gain*. TCM achieves these coding gains without requiring more bandwidth than the uncoded QPSK system.

As discussed in Chapter 3, a trellis code is generated by a circuit with a finite number of internal states. The number of these states is a direct measure of its decoding complexity if maximum-likelihood decoding is used. Note that the two very large codes are not maximum-likelihood decoded, but sequentially decoded [20]. Coding, then, partly realizes the promise by Shannon's theory, which states that for a desired error rate of $P_b = 10^{-6}$ we can gain almost 9 dB in expended signal energy over QPSK. This gain can be achieved by converting required signal power into required decoder complexity, as is done by the TCM methods.

Incidentally, 1948 is also the birth year [2] of the transistor, probably the most fundamental invention in this century, one that allowed the construction of very powerful, very small computers. Only this made the conversion from signal energy requirements to (circuit) complexity possible, giving coding and information theory a platform for practical realizations [1].

In Figure 1.3 we compare the performance of a 16-state 8-PSK TCM code used in an experimental implementation of a single-channel-per-

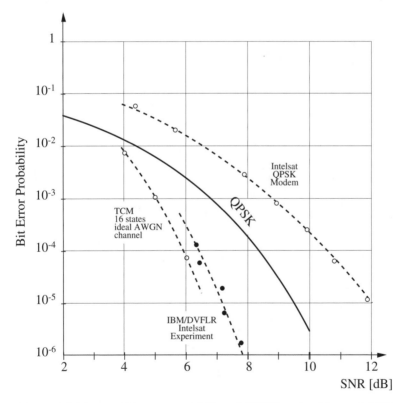

Figure 1.3 Measured bit error probability of (QPSK) and a 16-state 8-PSK (TCM) modem over a 64-kbit/s satellite channel [3].

[2] Transistor action was first observed on December 15, 1947, but the news of the invention was not made public until June 30, 1948.

carrier (SCPC) modem operating at 64 kbit/s [3] against QPSK and the theoretical performance established via simulations (Figure 1.2). This illustrates the viability of trellis coding for satellite channels. In fact, the 8-PSK TCM modem comes much closer to the theoretical performance than the original QPSK modem, and a coding gain of 5 dB is achieved.

Figure 1.4 shows the performance of selected convolutional codes on an additive white Gaussian noise channel (see also [9, 21–24]). Contrary to TCM, convolutional codes do not preserve bandwidth, and the gains in power efficiency in Figure 1.4 are partly obtained by a power bandwidth trade-off; i.e., the rate 1/2 convolutional codes require twice as much bandwidth as uncoded transmission, and rate 1/3 convolutional codes require three times as much. This bandwidth expansion may not be an issue in deep-space communications and the application of error control to spread-

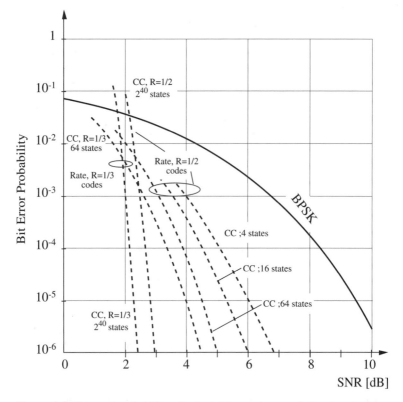

Figure 1.4 Bit error probability of selected low rate convolutional codes as a function of the normalized signal-to-noise ratio. The two very large codes are decoded sequentially, and the performance of all the other codes is for maximum-likelihood decoding. Simulation results are taken from [21] and [23].

spectrum systems [25, 26]. As a consequence, for the same complexity, convolutional codes achieve a higher coding gain than TCM. Nevertheless, a point of diminishing returns is achieved relatively quickly, and the cutoff rate of a channel, rather than its capacity, is often seen as the practical limit [11] (Chapter 3).

For very low target error probabilities, a tandem coding method, called *concatenated coding*, has become very popular [9, 27, 28]. In concatenated coding, as illustrated in Figure 1.5, the FEC codec is broken up into inner and outer codes. The inner code is most often a trellis code that performs the channel error control. The outer code is typically a high-rate Reed-Solomon (RS) block code, whose function is to clean up the residual output error of the inner code. This combination has proven very powerful, since the error mechanism of the inner code is well matched to the error-correcting capabilities of the outer system. Via this tandem construction, very low error rates are achievable.

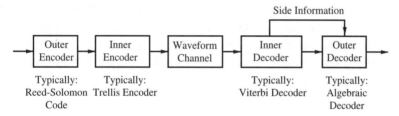

Figure 1.5 Concatenated FEC coding system using an inner codec and an outer codec.

The field of error control and error correction coding somewhat naturally breaks into two disciplines, block coding and trellis coding. Block coding, which is mostly approached as applied mathematics, has produced the bulk of publications in error control, yet trellis coding seems to be favored in most practical applications. One reason for this is surely the ease with which soft-decision decoding can be implemented for trellis codes. Soft decision is the operation when the demodulator no longer makes any (hard) decisions on the transmitted symbols but passes the received signal values directly onto the decoder. The decoder, in turn, operates on reliability information obtained by comparing the received signals with the possible set of transmitted signals. This gives soft decision a 2-dB head start. Also, trellis codes seem to be better matched to high-noise channels; that is, their performance is less sensitive to S/N variations than the performance of block codes. In many applications the trellis codes act as "S/N transformers" (e.g., [29]), improving the channel behavior as in concatenated coding. A mathematical curiosity is that trellis coding is much more difficult to analyze; consequently, scientists and engineers have been more willing to employ heuristic methods when dealing with trellis codes. The resulting codes and decoders, however, have proven to be very powerful, often outperforming the more structured block coding solutions. Even block codes can successfully be decoded by using decoding methods originally developed for trellis codes (Chapter 7).

1.3   BANDWIDTH AND POWER

Nyquist showed in 1928 [30] that a channel of bandwidth W (in Hz) is capable of supporting pulse-amplitude-modulated (PAM) signals at a rate of $2W$ samples/s without causing intersymbol interference. In other words, there are approximately $2W$ independent signal dimensions per second. If two carriers ($\sin(2\pi f_c)$ and $\cos(2\pi f_c)$) are used in quadrature, as in double-sideband suppressed carrier amplitude modulation (DSB-SC), we have W pairs of dimensions (or complex dimensions) per second, and the ubiquitous quadrature-amplitude-modulated (QAM) formats are born (Chapter 2).

The parameter that characterizes how efficiently a system uses its allotted bandwidth is the bandwidth efficiency η, defined as

$$\eta = \frac{\text{Bit rate}}{\text{Channel bandwidth } W} \quad \text{[bits/s/Hz].} \tag{1.2}$$

In order to compare different communications systems, a second parameter, expressing the power efficiency, has to be considered. This parameter is the information bit error probability P_b. For practical systems usually $10^{-3} \geq P_b \geq 10^{-5}$, though P_b is sometimes lower as for digital TV.

Using (1.1) and dividing by W, we obtain the maximum bandwidth efficiency for an additive white Gaussian noise channel, the *Shannon limit*, as

$$\eta_{\max} = \log_2\left(1 + \frac{S}{N}\right) \quad \text{[bits/s/Hz].} \tag{1.3}$$

To calculate η, we must suitably define W. This is obvious for some signaling schemes, such as Nyquist signaling, that have a rather sharply defined bandwidth (see Chapter 2), but becomes more arbitrary for modulation schemes with infinite spectral occupancy. One commonly used definition is the 99% bandwidth definition; i.e., W is defined such that 99% of the transmitted signal power falls within the band of width W. This 99% bandwidth corresponds to an out-of-band power of -20 dB.

Further, note that the average signal power S can be expressed as

$$S = \frac{kE_b}{T} = RE_b,$$ (1.4)

where E_b is the energy per bit, k is the number of bits transmitted per symbol, and T is the duration of that symbol. The parameter $R = k/T$ is the transmission rate of the system in bits/s (also called the *spectral bit rate*), as mentioned earlier. Rewriting the signal-to-noise power ratio S/N, where $N = N_0 W$—that is, total noise power equals the one-sided noise power spectral density (N_0) multiplied by the width of the transmission band—we obtain the Shannon limit in terms of the bit energy and noise power spectral density:

$$\eta_{max} = \log_2 \left(1 + \frac{RE_b}{N_0 W}\right).$$ (1.5)

Since $R/W = \eta_{max}$ is the limiting spectral efficiency, we obtain a bound from (1.5) on the minimum bit energy required for reliable transmission at a given spectral efficiency:

$$\frac{E_b}{N_0} \geq \frac{2^{\eta_{max}} - 1}{\eta_{max}},$$ (1.6)

also called the *Shannon bound*. This is the bound plotted in Figure 1.2 for a spectral efficiency of $\eta_{max} = 2$ bits/s/Hz and in Figure 1.6 as a function of E_b/N_0.

If spectral efficiency is not at a premium, and a large amount of bandwidth is available for transmission, we may choose to use bandwidth rather than power to increase the channel capacity (1.1). In the limit as the signal is allowed to occupy an infinite amount of bandwidth, that is, $\eta_{max} \to 0$, we obtain

$$\frac{E_b}{N_0} \geq \lim_{\eta_{max} \to 0} \frac{2^{\eta_{max}} - 1}{\eta_{max}} = \ln(2),$$ (1.7)

which is the minimum bit energy to noise power spectral density required for reliable transmission. This minimum $E_b/N_0 = \ln(2) = -1.59$ dB.

Figure 1.6 shows the power and bandwidth efficiencies of some popular uncoded quadrature constellations as well as that of a number of coded transmission schemes. The plot clearly demonstrates the benefits of coding. Trellis-coded modulation schemes used in practice, for example, achieve a power gain of up to 6 dB without loss in spectral efficiency. The convolutionally encoded methods (the points (2,1,7) QPSK, which is a rate $R = 1/2$ convolutional code with 2^7 states, and (4,1,14) QPSK, a rate

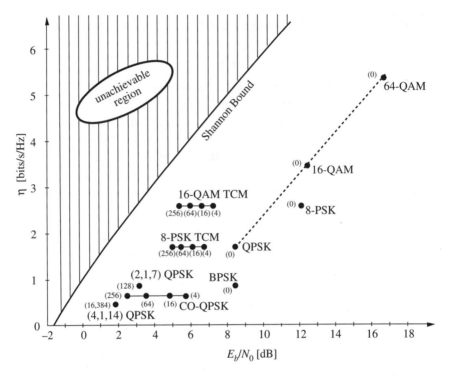

Figure 1.6 Spectral efficiencies achieved by various coded and uncoded transmission methods using spectrally raised cosine pulses with roll-off factor $\beta = 0.3$ (except CO-QPSK), bit error rate $P_b = 10^{-4}$, and $\eta = (1 + \beta)/T$. η for OC-QPSK is calculated from the 99% bandwidth.

$R = 1/4$ convolutional code with 2^{14} states) achieve a gain in power efficiency, but at the expense of spectral efficiency with respect to the original signal constellation. This is traditional bandwidth-expansion coding, and it was erroneously believed for quite some time that error control coding would always incur such a loss in bandwidth efficiency.

Coded overlapped quadrature modulation (CO-QPSK) is a combined coding and controlled intersymbol interference method [31] with smaller amplitude fluctuations than Nyquist signaling, and is a candidate for systems with amplifier nonlinearities, such as the satellite traveling-wave tube (TWT) amplifiers [32]. Yet another method, continuous-phase modulation (CPM), is a constant-amplitude modulation format [33] with similarities to TCM. CPM derives from frequency modulation and aims at improving the bandwidth efficiency by smoothing phase transitions between symbols. CPM has undergone an evolution similar to TCM, but despite its importance we will not treat CPM in this book. The reader is referred to the excellent book by Anderson, Aulin, and Sundberg [33], the standard reference for CPM.

It can be seen that all of these coding systems approximately fall into the same diagonal area of performance. The closer we get to the Shannon limit, the more complexity has to be expended in achieving the desired point. Even the most complex schemes are still about 2 dB away from the Shannon limit. Note also that for smaller target error probabilities this discrepancy increases, requiring even larger decoder complexity.

From basic communication theory we know that a signal of duration T and bandwidth occupancy W spans approximately $2.4WT$ orthogonal dimensions [34]. Since a finite-duration signal theoretically has an infinite bandwidth occupancy, the number of dimensions is somewhat difficult to describe precisely. For that reason, one often expresses Shannon's capacity formula (1.1) in terms of maximum rate per dimension. That is, if R_d is the rate in bits/dimension, then the capacity of an AWGN channel per dimension is the maximum rate at which reliable transmission is possible. It is given by [34]

$$C_d = \frac{1}{2} \log_2 \left(1 + 2\frac{R_d E_b}{N_0} \right) \quad \text{[bits/dimension]}. \qquad (1.8)$$

Applying the same manipulations as before, we obtain the Shannon bound normalized per dimension as

$$\frac{E_b}{N_0} \geq \frac{2^{2C_d} - 1}{2C_d}. \qquad (1.9)$$

Equation (1.9) is useful when the question of waveforms and pulse shaping is not a central issue, since it allows one to eliminate these considerations by treating signal dimensions rather than the signal itself (see also Chapter 2). We will use (1.9) for our comparisons in the next section.

1.4 HISTORY (THE DRIVE TOWARD CAPACITY)

Trellis coding celebrated its first success in the application of convolutional codes to deep-space probes in the 1960s and 1970s. For a long time afterward, error control coding was considered a curiosity with deep-space communications as its only viable application. This is the power-limited case and a picture-book success story of error control coding.

If we start with uncoded binary phase-shift keying (BPSK) as our baseline transmission method (see Chapter 2) and assume coherent detection, we can achieve a bit error rate of $P_b = 10^{-5}$ at a bit energy to noise power spectral density ratio of $E_b/N_0 = 9.6$ dB and a spectral efficiency of 1 bit/dimension. From the Shannon limit in Figure 1.7 it can be seen that 1 bit/dimension is theoretically achievable with $E_b/N_0 = 1.76$ dB, indicating that a power savings of nearly 8 dB is theoretically possible by applying coding.

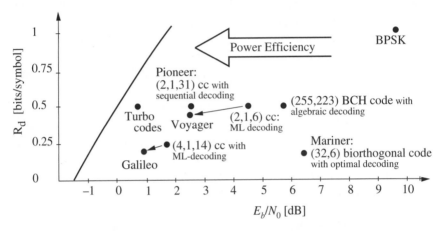

Figure 1.7 Milestones in the drive toward channel capacity achieved by the space systems which evolved over the past 40 years as an answer to the Shannon capacity challenge.

One of the earliest attempts to close this signal energy gap was the use of a rate 6/32 biorthogonal (Reed-Muller) block code [19]. This code was used on the *Mariner* Mars and Viking missions in conjunction with BPSK and soft-decision maximum-likelihood decoding. This system had a spectral efficiency of 0.1875 bit/symbol and achieved $P_b = 10^{-5}$ with $E_b/N_0 = 6.4$ dB. Thus, the (32,6) biorthogonal code required 3.2 dB less power than BPSK at the cost of a fivefold increase in the bandwidth. The performance of the 6/32 biorthogonal code is plotted in Figure 1.7, as are all the other systems discussed here.

In 1967, a new algebraic decoding technique was discovered for Bose-Chaudhuri-Hocquenghem (BCH) codes [35, 36], which enabled the efficient hard-decision decoding of an entire class of block codes. For example, the (255,123) BCH code has an $R_d \approx 0.5$ bit/symbol and achieves $P_b = 10^{-5}$ with $E_b/N_0 = 5.7$ dB using algebraic decoding.

The next step was taken with the introduction of sequential decoding (see Chapter 6), which could make use of soft-decision decoding. Sequential decoding allowed the decoding of long-constraint-length convolutional codes, and was first used on the *Pioneer 9* mission [37]. The *Pioneer 10* and

Pioneer 11 missions in 1972 and 1973 both used a long-constraint-length (2,1,31), nonsystematic convolutional code (Chapter 4) [38]. A sequential decoder was used that achieved $P_b = 10^{-5}$ with $E_b/N_0 = 2.5$ dB and $R_d = 0.5$. This is only 2.5 dB away from the capacity of the channel.

Sequential decoding has the disadvantage that the computational load is variable, and this load grows exponentially the closer the operation point moves toward capacity (see Chapter 6). For this and other reasons, the next generation of space systems employed maximum-likelihood decoding. The *Voyager* spacecraft launched in 1977 used a short-constraint-length (2,1,6) convolutional code in conjunction with a soft-decision Viterbi decoder achieving $P_b = 10^{-5}$ at $E_b/N_0 = 4.5$ dB and a spectral efficiency of $R_d = 0.5$ bit/symbol. The biggest Viterbi decoder built to date [39] found application in the *Galileo* mission, where a (4,1,14) convolutional code was used. This yields a spectral efficiency of $R_d = 0.25$ bit/symbol and achieves $P_b = 10^{-5}$ at $E_b/N_0 = 1.75$ dB. The performance of this system is also 2.5 dB away from the capacity limit. The systems for *Voyager* and *Galileo* are further enhanced by the use of concatenation in addition to the convolutional inner code. An outer (255,223) Reed-Solomon code [19] is used to reduce the required signal-to-noise ratio by 2.0 dB for the *Voyager* system and by 0.8 dB for the *Galileo* system.

More recently, Turbo codes [40] using iterative decoding have virtually closed the gap to capacity by achieving $P_b = 10^{-5}$ at a spectacularly low E_b/N_0 of 0.7 dB with $R_d = 0.5$ bit/symbol. It appears that the 40-year effort to reach capacity has been achieved with this latest invention. Turbo codes are treated in detail in Chapter 8.

Space applications of error control coding have met with spectacular success, and for a long time the belief that coding was useful only in improving power efficiency of digital transmission was prevalent. This attitude was thoroughly overturned by the spectacular success of error control coding on voiceband data transmission systems. Here it was not the power efficiency but the spectral efficiency that was the issue; that is, given a standard telephone channel with an essentially fixed bandwidth and SNR, what was the maximum practical rate of reliable transmission?

The first commercially available voiceband modem in 1962 achieved a transmission rate of 2400 bit/s. Over the next 10 to 15 years these rates improved to 9600 bits/s, which was then considered to be the maximum achievable rate, and efforts to push the rate higher were frustrated. Ungerböck's invention of TCM in the late 1970s, however, opened the door to further, unexpected improvements. The modem rates jumped to 14,400 bits/s and then to 19,200 bits/s, using sophisticated TCM schemes

[41]. The latest chapter in voiceband data modems is the establishment of the CCITT V.34 modem standard [42]. The modems specified therein achieve a maximum transmission rate of 28,800 bits/s, and extensions to V.34 to cover two new rates at 31,200 bits/s and 33,600 bits/s have been proposed. However, at these high rates modems operate success-fully only on a small percentage of the connections. It seems that the limits of the voiceband telephone channel have been reached (accord-ing to [43]). This needs to be compared to estimates of the channel ca-pacity for a voiceband telephone channel, which are somewhere around 30,000 bits/s. The application of TCM is one of the fastest migrations of an experimental laboratory system to an international standard (V.32–V.34) [42, 44]. Hence, what once was believed to be impossible, achiev-ing capacity on band-limited channels, has become commonplace reality within about a decade and a half after the invention of TCM [12]. The trellis codes used in these advanced modems are discussed in detail in Chapter 3.

In many ways the telephone voiceband channel was an ideal playground for the application of error control coding. Its limited bandwidth of about 3 kHz (400–3400 Hz) implies relatively low data rates by modern standards. It therefore provides an ideal experimental field for high-complexity error control methods, which can be implemented without much difficulty using current digital signal processing (DSP) technology. It is thus not surprising that coding for voiceband channels was the first successful application of bandwidth-efficient error control.

Nowadays, trellis coding in the form of bandwidth-efficient TCM as well as more conventional convolutional coding is being considered for satellite communications, both geostationary and low-earth-orbiting satellites, for land-mobile and satellite-mobile services, cellular commu-nications networks, personal communications services (PCS), and high-frequency (HF) tropospheric long-range communications, among others. It seems that the age of widespread application of error control coding has only just begun.

REFERENCES

[1] H. Imai et al., *Essentials of Error-Control Coding Techniques*, Academic Press, New York, 1990.

[2] S. A. Rhodes, R. J. Fang, and P. Y. Chang, "Coded octal phase shift key-ing in TDMA satellite communications," *COMSAT Tech. Rev.*, Vol. 13, pp. 221–258, 1983.

[3] G. Ungerböck, J. Hagenauer, and T. Abdel-Nabi, "Coded 8-PSK experimental modem for the INTELSAT SCPC system," *Proc. ICDSC, 7th*, pp. 299–304, 1986.

[4] M. Mouly and M.-B. Pautet, *The GSM System for Mobile Communications*, 1993.

[5] TIA/EIA/IS-95 interim standard, mobile station–base station compatibility standard for dual-mode wideband spread spectrum cellular systems, Telecommunications Industry Association, Washington, DC, July 1993.

[6] N. S. Jayant and P. Noll, *Digital Coding of Waveforms*, Prentice Hall, Englewood Cliffs, NJ, 1984.

[7] H. Meyr and G. Ascheid, *Synchronization in Digital Communications*, Vol. 1, Wiley, New York, 1990.

[8] S. Lin, D. J. Costello, Jr., and M. J. Miller, "Automatic-repeat-request error-control schemes," *IEEE Commun. Mag.*, Vol. 22, pp. 5–17, 1984.

[9] S. Lin and D. J. Costello, Jr., *Error Control Coding: Fundamentals and Applications*, Prentice Hall, Englewood Cliffs, NJ, 1983.

[10] J. M. Wozencraft and R. S. Kennedy, "Modulation and demodulation for probabilistic coding," *IEEE Trans. Inform. Theory*, Vol. IT-12, No. 3, pp. 291–297, 1966.

[11] J. L. Massey, "Coding and modulation in digital communications," *Proc. Int. Zürich Sem. Digital Commun.*, Zürich, Switzerland, pp. E2(1)–E2(4), March 1974.

[12] G. Ungerböck, "Channel coding with multilevel/phase signals," *IEEE Trans. Inform. Theory*, Vol. 28, No. 1, pp. 55–67, 1982.

[13] G. Ungerböck, "Trellis-coded modulation with redundant signal sets. I: Introduction," *IEEE Commun. Mag.*, Vol. 25, No. 2, pp. 5–11, 1987.

[14] G. Ungerböck, "Trellis-coded modulation with redundant signal sets. II: State of the art," *IEEE Commun. Mag.*, Vol. 25, No. 2, pp. 12–21, 1987.

[15] J. Hagenauer, "Rate compatible punctured convolutional codes (RCPC-codes) and their application," *IEEE Trans. Commun.*, Vol. COM-36, pp. 389–400, 1988.

[16] C. E. Shannon, "A mathematical theory of communications," *Bell Syst. Tech. J.*, Vol. 27, pp. 379–423, 1948.

[17] R. W. Hamming, "Error detecting and error correcting codes," *Bell Syst. Tech. J.*, Vol. 29, pp. 147–160, 1950.

[18] M. J. E. Golay, "Notes on digital coding," *Proc. IEEE*, Vol. 37, p. 657, 1949.

[19] F. J. MacWilliams and N. J. A. Sloane, *The Theory of Error Correcting Codes*, North-Holland, Amsterdam, 1988.

[20] F.-Q. Wang and D. J. Costello, "Probabilistic construction of large constraint length trellis codes for sequential decoding," *IEEE Trans. Commun.*, Vol. COM-43, pp. 2439–2448, 1995.

[21] J. A. Heller and J. M. Jacobs, "Viterbi detection for satellite and space communications," *IEEE Trans. Commun. Technol.*, Vol. COM-19, pp. 835–848, 1971.

[22] J. P. Odenwalder, "Optimal decoding of convolutional codes," Ph.D. thesis, University of California, Los Angeles, 1970.

[23] J. K. Omura and B. K. Levitt, "Coded error probability evaluation for anti-jam communication systems," *IEEE Trans. Commun.*, Vol. 30, pp. 896–903, 1982.

[24] J. G. Proakis, *Digital Communications,* McGraw-Hill, New York, 1989.

[25] A. J. Viterbi, "Spread spectrum communications—myths and realities," *IEEE Commun. Mag.*, Vol. 17, pp. 11–18, 1979.

[26] A. J. Viterbi, "When not to spread spectrum—a sequel," *IEEE Commun. Mag.*, Vol. 23, pp. 12–17, 1985.

[27] G. D. Forney, *Concatenated Codes*, MIT Press, Cambridge, MA, 1966.

[28] K. Y. Lin and J. Lee, "Recent results on the use of concatenated Reed-Solomon/Viterbi channel coding and data compression for space communications," *IEEE Trans. Commun.*, Vol. 32, pp. 518–523, 1984.

[29] J. Hagenauer and P. Höher, "A Viterbi algorithm with soft-decision outputs and its applications," *Proc. IEEE Globecom'89*, 1989.

[30] H. Nyquist, "Certain topics in telegraph transmission theory," *AIEE Trans.*, pp. 617ff., 1946.

[31] C. Schlegel, "Coded overlapped quadrature modulation," *Proc. Global Conf. Commun. GLOBECOM'91*, Phoenix, AZ, December 1991.

[32] S. Ramseier and C. Schlegel, "On the bandwidth/power tradeoff of trellis coded modulation schemes," *Proc. IEEE Globecom'93*, 1993.

[33] J. B. Anderson, T. Aulin, and C-E. Sundberg, *Digital Phase Modulation*, Plenum Press, New York, 1986.

[34] J. M. Wozencraft and I. M. Jacobs, *Principles of Communication Engineering*, Wiley, New York, 1965; reprinted by Waveland Press, 1993.

[35] E. R. Berlekamp, *Algebraic Coding-Theory*, McGraw-Hill, New York, 1968.

[36] J. L. Massey, "Shift register synthesis and BCH decoding," *IEEE Trans. Inform. Theory*, Vol. IT-15, pp. 122–127, 1969.

[37] G. D. Forney, "Final report on a study of a sample sequential decoder," Appendix A, Codex Corp., Watertown, MA, U.S. Army Satellite Communication Agency Contract DAA B 07-68-C-0093, April 1968.

[38] J. L. Massey and D. J. Costello, Jr., "Nonsystematic convolutional codes for sequential decoding in space applications," *IEEE Trans. Commun.*, Vol. COM-19, pp. 806–813, 1971.

[39] O. M. Collins, "The subtleties and intricacies of building a constraint length 15 convolutional decoder," *IEEE Trans. Commun.*, Vol. COM-40, pp. 1810–1819, 1992.

[40] C. Berrou, A. Glavieux, and P. Thitimajshima, "Near Shannon limit error-correcting coding and decoding: Turbo-codes," *Proc. 1993 IEEE Int. Conf. on Comm.*, Geneva, Switzerland, pp. 1064–1070, 1993.

[41] G. D. Forney, "Coded modulation for bandlimited channels," *IEEE Information Theory Society Newsletter*, December 1990.

[42] CCITT Recommendations V.34.

[43] R. Wilson, "Outer limits," *Electronic News*, May 1996.

[44] U. Black, *The V Series Recommendations, Protocols for Data Communications Over the Telephone Network*, McGraw-Hill, New York, 1991.

2

COMMUNICATION
THEORY BASICS

2.1 THE PROBABILISTIC VIEWPOINT

Communication, the transmission of information from a sender to a receiver (destination), is basically a random experiment. The sender selects one of a number of possible messages which is presented (transmitted) to the receiver. The receiver has no knowledge of which message is chosen by the sender, for if it did there would be no need for transmission. The transmitted message is chosen from a set of messages known to the sender and the receiver. In light of this, it should not be surprising us that many quantities in modern communication theory are random quantities.

Figure 2.1 shows a simplified system block diagram of such a sender/receiver communication system. The transmitter performs the random experiment of selecting one of the M messages in the message set $\{m^{(i)}\}$, say $m^{(i)}$, and transmits a corresponding signal $s^{(i)}(t)$ chosen from a set of signals $\{s^{(i)}(t)\}$. In many cases this signal is a continuous function, particularly in the cases with which this book is concerned. The transmission medium is called the channel and may be a telephone wire line, a radio link, an underwater acoustic link or any other suitable arrangement, as elaborated in Chapter 1. One of the major impairments in all communication systems is noise, most importantly, thermal noise. Thermal noise is generated by the random motion of particles inside the receiver's signal-sensing elements and has the property that it adds linearly to the received signal, hence the channel model in Figure 2.1.

A large body of literature is available on modeling channel impairments on the transmitted signal. In this book we concentrate on the almost ubiquitous case of additive white Gaussian noise (AWGN), which is reviewed briefly in Appendix 2A. Noise is itself a random quantity, and our

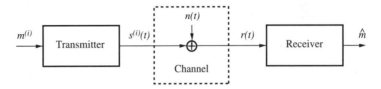

Figure 2.1 Block diagram of a sender/receiver communication system used
on a channel with additive noise.

communication system becomes a joint random experiment. The output
of the transmitter and the output of the noise source are both random quantities. More precisely, using Figure 2.1, these quantities are random processes (i.e., random functions). It is now the task of the receiver to estimate the selected message $m^{(i)}$ with the highest achievable reliability. In
the context of a discrete set of messages, one speaks of message detection,
rather than estimation, since the receiver is not trying to reproduce closely
a given signal, but makes a decision for one of a finite number of messages. As a measure of the reliability of this process we use the almost
universally accepted probability of error, which is defined as the probability that the message $\hat{m}$ identified by the receiver is not the one originally
chosen by the transmitter; that is, $P_e = \Pr(\hat{m} \neq m^{(i)})$. The solution to this
problem is essentially completely known for many cases, and we will subsequently present an overview of the optimal receiver principles in additive
Gaussian noise.

2.2   VECTOR COMMUNICATION CHANNELS

A very popular way to generate the signals at the transmitter is to synthesize
them as a linear combination of N *basis waveforms* $\phi_j(t)$; that is, the
transmitter selects

$$s^{(i)}(t) = \sum_{j=1}^{N} s_j^{(i)} \phi_j(t) \qquad (2.1)$$

as the transmitted signal for the ith message. Often the basis waveforms
are chosen to be orthonormal; that is, they fulfill the condition

$$\int_{-\infty}^{\infty} \phi_j(t)\phi_l(t)\,dt = \delta_{jl} = \begin{cases} 1, & j = l, \\ 0, & \text{otherwise.} \end{cases} \qquad (2.2)$$

This leads to a vector interpretation of the transmitted signals, since, once
the basis waveforms are specified, $s^{(i)}(t)$ is completely determined by the

N-dimensional vector

$$\mathbf{s}^{(i)} = \left(s_1^{(i)}, s_2^{(i)}, \ldots, s_N^{(i)}\right). \tag{2.3}$$

We can now visualize the signals geometrically by viewing the signal vectors $\mathbf{s}^{(i)}$ in Euclidean N-space, spanned by the usual orthonormal basis vectors, where each basis vector is associated with a basis function. This geometric representation of signals is called a signal constellation. The idea is illustrated for $N = 2$ in Figure 2.2 for the signals $s^{(1)}(t) = \sin(2\pi f_1 t)w(t)$, $s^{(2)}(t) = \cos(2\pi f_1 t)w(t)$, $s^{(3)}(t) = -\sin(2\pi f_1 t)w(t)$, and $s^{(4)}(t) = -\cos(2\pi f_1 t)w(t)$, where

$$w(t) = \begin{cases} \sqrt{2E_s/T}, & 0 \le t \le T, \\ 0, & \text{otherwise}, \end{cases} \tag{2.4}$$

and $f_1 = k/T$ is an integer multiple of $1/T$. The first basis function is $\phi_1(t) = \sqrt{2/T}\sin(2\pi f_1 t)$, and the second basis function is $\phi_2(t) = \sqrt{2/T}\cos(2\pi f_1 t)$.

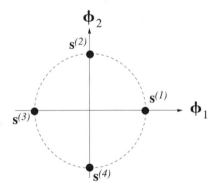

Figure 2.2 Illustration of four signals using two orthogonal basis functions (QPSK).

The signal constellation in Figure 2.2 is called quadrature phase-shift keying (QPSK). Note that while we have lost the information on the actual waveform, we have gained a higher level of abstraction, which will make it much easier to discuss subsequent concepts of coding and modulation.

Knowledge of the signal vector $\mathbf{s}^{(i)}$ implies knowledge of the transmitted message $m^{(i)}$, since, in a sensible system, there is a one-to-one mapping between the two. The problem of decoding a received waveform is therefore equivalent to recovering the signal vector $\mathbf{s}^{(i)}$. This can be accomplished by passing the received signal waveform through a bank of correlators where each correlates $s^{(i)}(t)$ with one of the basis functions, performing the operation

$$\int_{-\infty}^{\infty} s^{(i)}(t)\phi_l(t)\, dt = \sum_{j=1}^{N} s_j^{(i)} \int_{-\infty}^{\infty} \phi_j(t)\phi_l(t)\, dt = s_l^{(i)}; \qquad (2.5)$$

that is, the lth correlator recovers the lth component $s_l^{(i)}$ of the signal vector $\mathbf{s}^{(i)}$.

Later we will need the *squared Euclidean distance* between two signals $\mathbf{s}^{(i)}$ and $\mathbf{s}^{(j)}$, given by $d_{ij}^2 = |\mathbf{s}^{(i)} - \mathbf{s}^{(j)}|^2$, which is a measure of the noise resistance of these two signals. Furthermore,

$$
\begin{aligned}
d_{ij}^2 &= \sum_{l=1}^{N} \left(s_l^{(i)} - s_l^{(j)} \right)^2 \\
&= \sum_{l=1}^{N} \left(\int_{-\infty}^{\infty} \left(s_l^{(i)} - s_l^{(j)} \right) \phi_l(t) \right)^2 \qquad (2.6) \\
&= \int_{-\infty}^{\infty} \left(s^{(i)}(t) - s^{(j)}(t) \right)^2 dt
\end{aligned}
$$

is in fact the energy of the difference signal $s^{(i)}(t) - s^{(j)}(t)$. (It is easier to derive (2.6) in the reverse direction, i.e., from bottom to top.)

It can be shown[1] that this correlator receiver is optimal, in the sense that no relevant information is discarded and minimum error probability can still be attained, even when the received signal contains additive white Gaussian noise. In this case the received signal $r(t) = s^{(i)}(t) + n_w(t)$ produces the received vector $\mathbf{r} = \mathbf{s}^{(i)} + \mathbf{n}$ at the output of the bank of correlators. The statistics of the noise vector $\mathbf{n}$ can easily be evaluated, using the orthogonality of the basis waveforms and the noise correlation function $E[n_w(t)n_w(t + \tau)] = \delta(\tau)N_0/2$, where $\delta(t)$ is *Dirac's delta function* and N_0 is the *one-sided noise power spectral density* (see Appendix 2.A). We then obtain

$$
\begin{aligned}
E\left[n_l n_j \right] &= \int_{-\infty}^{\infty} \int_{-\infty}^{\infty} E\left[n_w(\alpha)n_w(\beta) \right] \phi_l(\alpha)\phi_j(\beta)\, d\alpha\, d\beta \\
&= \frac{N_0}{2} \int_{-\infty}^{\infty} \phi_l(\alpha)\phi_j(\alpha)\, d\alpha = \begin{cases} \dfrac{N_0}{2}, & l = j, \\ 0, & \text{otherwise.} \end{cases}
\end{aligned}
\qquad (2.7)
$$

We are pleased to see that, courtesy of the orthonormal basis waveforms, the components of the random noise vector $\mathbf{n}$ are all uncorrelated. Since $n_w(t)$

---

[1] The book by Wozencraft and Jacobs [1] contains an excellent treatise on finite-dimensional signal spaces.

is a Gaussian random process, the sample values n_l, n_j are necessarily also Gaussian. (For more detail see, e.g., [2, 3].) From the foregoing we conclude that the components of $\mathbf{n}$ are independent, Gaussian random variables with common variance $N_0/2$ and mean zero.

We have thus achieved a completely equivalent vector view of a communications system. The advantages of this point of view are manifold. First, we need not concern ourselves with the actual choices of signal waveforms when discussing receiver algorithms, for example; second, the difficult problem of waveform communication involving stochastic processes and continuous signal functions has been transformed into the much more manageable vector communications system involving only random vectors and signal vectors. We have thus gained a geometric view of communications. In this context, linear algebra, the tool for geometric operations, plays an important role in modern communications theory.

The vector representation is finite dimensional, the number of dimensions given by the dimensionality of the signal functions. The random function space is infinite dimensional, however. The optimality of the correlator receiver shows then that we need only be concerned with that part of the noise projected onto the signal space. All other noise components are irrelevant. In fact, this is the way the optimality of the correlator receiver is typically proven [1]: First it is shown that the noise components that are not part of the finite-dimensional signal space are irrelevant and need not be considered by the receiver. This leads directly to the finite-dimensional geometric signal space interpretation discussed before.

2.3 OPTIMAL RECEIVERS

If our bank of correlators produces a received vector $\mathbf{r} = \mathbf{s}^{(i)} + \mathbf{n}$, then an optimal detector will choose as message hypothesis, $\hat{m} = m^{(j)}$, the one that maximizes the conditional probability

$$P[m^{(j)}|\mathbf{r}], \qquad (2.8)$$

because this maximizes the overall probability of being correct, P_c, which can be seen from

$$P_c = \int_{\mathbf{r}} P[\text{correct}|\mathbf{r}] p(\mathbf{r}) \, d\mathbf{r}; \qquad (2.9)$$

that is, since $p(\mathbf{r})$ is positive, maximizing P_c can be achieved by maximizing $P[\text{correct}|\mathbf{r}]$ for each $\mathbf{r}$. This is precisely what maximizing (2.8) accomplishes. This is known as a maximum a posteriori (MAP) receiver.

Using Bayes' rule we obtain

$$P[m^{(j)}|\mathbf{r}] = \frac{P[m^{(j)}]p(\mathbf{r}|m^{(j)})}{p(\mathbf{r})}, \tag{2.10}$$

and, postulating that the signals are all used equally likely, it suffices for the receiver to select $\hat{m} = m^{(j)}$ such that $p(\mathbf{r}|m^{(j)})$ is maximized. This is the *maximum-likelihood* (ML) receiver. It minimizes the signal error probability only for equally likely signals. Thus, MAP and ML are equivalent if the signals are equally likely.

Since $\mathbf{r} = \mathbf{s}^{(i)} + \mathbf{n}$ and $\mathbf{n}$ is an additive Gaussian random vector independent of the signal $\mathbf{s}^{(i)}$, we may further develop the optimal receiver by using $p(\mathbf{r}|m^{(j)}) = p_n(\mathbf{r} - \mathbf{s}^{(j)})$, which is an N-dimensional Gaussian density function given by

$$p_n(\mathbf{r} - \mathbf{s}^{(j)}) = \frac{1}{(\pi N_0)^{N/2}} \exp\left(-\frac{|\mathbf{r} - \mathbf{s}^{(j)}|^2}{N_0}\right). \tag{2.11}$$

Maximizing (2.8) is now seen to be equivalent to minimizing the Euclidean distance

$$|\mathbf{r} - \mathbf{s}^{(j)}|^2 \tag{2.12}$$

between the received vector and the hypothesized signal vector $\mathbf{s}^{(j)}$.

The decision rule (2.12) implies decision regions $\mathcal{D}^{(j)}$ for each signal point that consists of all the points in Euclidean N-space that are closer to $\mathbf{s}^{(j)}$ than any other signal point. These regions are also known as *Voronoi regions*. The decision regions are illustrated for QPSK in Figure 2.3.

The probability of error given a particular transmitted signal $\mathbf{s}^{(i)}$ can now be interpreted as the probability that the additive noise $\mathbf{n}$ carries the

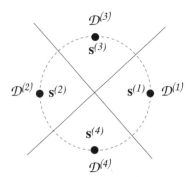

Figure 2.3 Decision regions for ML detection of equiprobable QPSK signals.

signal $\mathbf{s}^{(i)}$ outside its decision region $\mathcal{D}^{(i)}$. This probability can be calculated as

$$P_e(\mathbf{s}^{(i)}) = \int_{\mathbf{r} \notin \mathcal{D}^{(i)}} p_n(\mathbf{r} - \mathbf{s}^{(i)}) \, d\mathbf{r}. \qquad (2.13)$$

Equation (2.13) is, in general, very difficult to calculate, and simple expressions exist only for some special cases. The most important such special case is the two-signal error probability, which is defined as the probability that signal $\mathbf{s}^{(i)}$ is decoded as signal $\mathbf{s}^{(j)}$, assuming that there are only these two signals. To calculate the two-signal error probability we disregard all signals except $\mathbf{s}^{(i)}, \mathbf{s}^{(j)}$. Take, for example, $\mathbf{s}^{(i)} = \mathbf{s}^{(1)}$ and $\mathbf{s}^{(j)} = \mathbf{s}^{(3)}$ in Figure 2.3. The new decision regions are $\mathcal{D}^{(i)} = \mathcal{D}^{(1)} \cup \mathcal{D}^{(4)}$ and $\mathcal{D}^{(j)} = \mathcal{D}^{(2)} \cup \mathcal{D}^{(3)}$. The decision region $\mathcal{D}^{(i)}$ is expanded to a half-plane and the probability of deciding on message $m^{(j)}$ when message $m^{(i)}$ was actually transmitted, which is known as the *pairwise error probability*, is

$$P_{\mathbf{s}^{(i)} \to \mathbf{s}^{(j)}} = Q\left(\sqrt{\frac{d_{ij}^2}{2N_0}}\right), \qquad (2.14)$$

where $d_{ij}^2 = |\mathbf{s}^{(i)} - \mathbf{s}^{(j)}|^2$ is the energy of the difference signal, and

$$Q(\alpha) = \frac{1}{\sqrt{2\pi}} \int_\alpha^\infty \exp\left(-\frac{\beta^2}{2}\right) d\beta, \qquad (2.15)$$

is a nonelementary integral, referred to as the (Gaussian) Q-function. For a more detailed discussion, see [1]. In any case, equation (2.14) states that the probability of error between two signals decreases exponentially with their squared Euclidean distance d_{ij}, due to the overbound [1]

$$Q\left(\sqrt{d_{ij}^2/2N_0}\right) \leq (1/2) \exp\left(-d_{ij}^2/4N_0\right). \qquad (2.16)$$

2.4   MATCHED FILTERS

The correlation operation (2.5) used to recover the signal vector components can be implemented as a filtering operation. The signal $s^{(i)}(t)$ is passed through a filter with impulse response $\phi_l(-t)$ to obtain

$$u(t) = s^{(i)}(t) \star \phi_l(-t) = \int_{-\infty}^\infty s^{(i)}(\alpha)\phi_l(\alpha - t) \, d\alpha. \qquad (2.17)$$

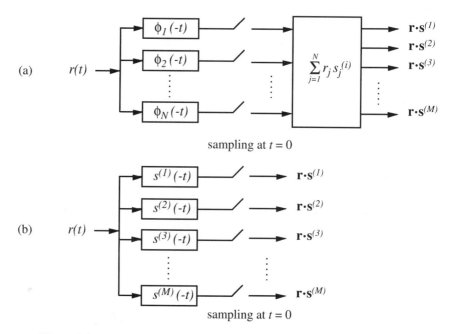

Figure 2.4 (a) The basis-function-matched filter receiver and (b) the signal-matched filter receiver. M is the number of messages.

If the output of the filter $\phi_l(-t)$ is sampled at time $t = 0$, equations (2.17) and (2.5) are identical; i.e.,

$$u(t = 0) = s_l^{(i)} = \int_{-\infty}^{\infty} s^{(i)}(\alpha)\phi_l(\alpha)\,d\alpha. \tag{2.18}$$

Of course, some appropriate delay needs to be built into the system to guarantee that $\phi_l(-t)$ is causal. We shall not be concerned with this delay in our treatise.

The maximum-likelihood receiver now minimizes $|\mathbf{r} - \mathbf{s}^{(i)}|^2$ or, equivalently, maximizes

$$2 \cdot \mathbf{r} \cdot \mathbf{s}^{(i)} - |\mathbf{s}^{(i)}|^2, \tag{2.19}$$

where we have neglected the term $|\mathbf{r}|^2$, which is common to all the hypotheses. The correlation $\mathbf{r} \cdot \mathbf{s}^{(i)}$ is the central part of (2.19) and can be implemented in two ways: as the basis-function-matched filter receiver, performing the summation after correlation—that is,

$$\mathbf{r} \cdot \mathbf{s}^{(i)} = \sum_{j=1}^{N} r_j s_j^{(i)} = \sum_{j=1}^{N} s_j^{(i)} \int_{-\infty}^{\infty} r(t)\phi_j(t)\,dt \tag{2.20}$$

—or as the signal-matched filter receiver performing

$$\mathbf{r} \cdot \mathbf{s}^{(i)} = \int_{-\infty}^{\infty} r(t) \sum_{j=1}^{N} s_j^{(i)} \phi_j(t) \, dt = \int_{-\infty}^{\infty} r(t) s^{(i)}(t) \, dt. \qquad (2.21)$$

The two receiver implementations are illustrated in Figure 2.4. Usually if the number of basis functions is much smaller than the number of signals, the basis-function-matched filter implementation is preferred.

The optimality of the correlation receiver (2.5) implies that both receiver structures of Figure 2.4 are optimal also. The sampled outputs of the matched filters are therefore sufficient for optimal detection of the transmitted message, and hence form what is known in statistics as a *sufficient statistics*.

2.5 MESSAGE SEQUENCES

In practice, our signals $s^{(i)}(t)$ will most often consist of a sequence of identical, time-displaced waveforms, called pulses, described by

$$s^{(i)}(t) = \sum_{r=-l}^{l} a_r p(t - rT), \qquad (2.22)$$

where $p(t)$ is some pulse waveform, the a_r are the discrete symbol values from some finite signal alphabet (e.g., binary signaling: $a_r \in \{-1, 1\}$), and $2l + 1$ is the length of the sequence in numbers of symbols. The parameter T is the timing delay between successive pulses, also called the *symbol period*. The output of the filter matched to the signal $s^{(i)}(t)$ is given by

$$y(t) = \int_{-\infty}^{\infty} r(\alpha) \sum_{r=-l}^{l} a_r p(\alpha - rT - t) \, d\alpha$$

$$= \sum_{r=-l}^{l} a_r \int_{-\infty}^{\infty} r(\alpha) p(\alpha - rT - t) \, d\alpha, \qquad (2.23)$$

and the sampled value of $y(t)$ at $t = 0$ is given by

$$y(t = 0) = \sum_{r=-l}^{l} a_r \int_{-\infty}^{\infty} r(\alpha) p(\alpha - rT) \, d\alpha = \sum_{r=-l}^{l} a_r y_r, \qquad (2.24)$$

where $y_r = \int_{-\infty}^{\infty} r(\alpha) p(\alpha - rT)$ is the output of a filter matched to the pulse $p(t)$ sampled at time $t = rT$. The signal-matched filter can therefore be implemented by a pulse-matched filter, whose output $y(t)$ is sampled at multiples of the symbol time T.

The time-shifted waveforms $p(t - rT)$ may serve as orthonormal basis functions if they fulfill the orthogonality condition

$$\int_{-\infty}^{\infty} p(t - rT)p(t - hT)\, dt = \delta_{rh}. \tag{2.25}$$

In any case, the output waveform of the pulse-matched filter $p(-t)$ is given by

$$z(t) = \sum_{r=-l}^{l} a_r p(t - rT) \star p(-t) = \sum_{r=-l}^{l} a_r g(t - rT), \tag{2.26}$$

where $g(t - rT) = \int_{-\infty}^{\infty} p(\alpha - rT)p(\alpha - t)\, d\alpha$ is the composite pulse/pulse-matched filter waveform. If (2.25) holds, $z(t)$ is completely separable and the rth sample value $z(rT) = z_r$ depends only on a_r, even if the pulses $p(t - rT)$ overlap; that is, $z_r = a_r$. If successive symbols a_r are chosen independently, the system may then be viewed as using a one-dimensional communication system independently $2l + 1$ times. This results in tremendous savings in complexity and is state-of-the-art signaling.

Condition (2.25) ensures that symbols transmitted at different times do not interfere with each other; that is, we have *intersymbol-interference-free signaling*. Condition (2.25) is known as the Nyquist criterion, which requires that the composite waveform $g(t)$ pass through zero at all multiples of the sampling time T, except $t = 0$; that is,

$$\int_{-\infty}^{\infty} p(t - rT)p(t - hT)\, dt = \delta_{rh} \Rightarrow g((h - r)T) = \begin{cases} 1, & \text{if } (h - r) = 0, \\ 0, & \text{otherwise.} \end{cases} \tag{2.27}$$

An example of such a composite pulse $g(t)$ together with its spectrum $G(f)$ is shown in Figure 2.5. Note that $G(f)$ also happens to be the

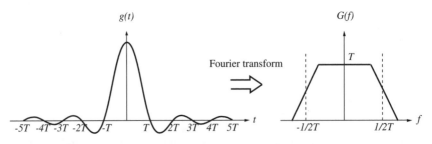

Figure 2.5 Example of a Nyquist pulse with trapezoid frequency spectrum.

energy spectrum of the transmitted pulse $p(t)$; that is, $G(f) = P(f)P^*(f) = |P(f)|^2$.

The Nyquist criterion can be translated into the frequency domain via the Fourier transform by observing that

$$
\begin{aligned}
g(rT) &= \int_{-\infty}^{\infty} G(f)e^{2\pi jfrT}\,df \\
&= \sum_{n=-\infty}^{\infty} \int_{(2n-1)/2T}^{(2n+1)/2T} G(f)e^{2\pi jfrT}\,df \\
&= \int_{-1/2T}^{1/2T} \sum_{n=-\infty}^{\infty} G\left(f + \frac{n}{T}\right) e^{2\pi jfrT}\,df.
\end{aligned}
\tag{2.28}
$$

If the folded spectrum

$$
\sum_{n=-\infty}^{\infty} G\left(f + \frac{n}{T}\right), \qquad -\frac{1}{2T} \le f \le \frac{1}{2T},
\tag{2.29}
$$

equals a constant, T for normalization reasons, the integral in (2.28) evaluates to $g(rT) = \sin(\pi r)/\pi r = \delta_{0r}$; that is, the sample values of $g(t)$ are zero at all nonzero multiples of T, as required. Equation (2.29) is the frequency spectral form of the Nyquist criterion for no intersymbol interference (compare Figure 2.5).

A very popular Nyquist pulse is the spectral raised cosine pulse whose spectrum is given by

$$
G(f) = \begin{cases}
T, & |f| \le (1-\beta)/2T, \\
T\cos^2\left(\dfrac{\pi}{4\beta}(2fT + \beta - 1)\right), & (1-\beta)/2T \le |f| \le (1+\beta)/2T, \\
0, & (1+\beta)/2T \le |f|,
\end{cases}
\tag{2.30}
$$

where $\beta \in [0, 1]$ is the roll-off factor of the pulse, whose bandwidth is $(1 + \beta)/2T$. In practical systems values of β around 0.3 are routinely used, and filters with β as low as 0.1 can be approximated by realizable filters, producing very bandwidth efficient signals.

These observations lead to the extremely important and popular root-Nyquist signaling method shown in Figure 2.6. The term stems from the fact that the actual pulse shape used, $p(t)$, is the inverse Fourier transform of $\sqrt{G(f)}$, the Nyquist pulse. In the case of a rectangular brick-wall frequency response, $p(t) = g(t) = \sin(\pi t/T)/(\pi t/T)$.

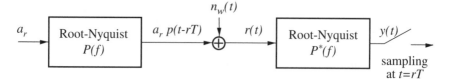

Figure 2.6 An optimal communication system using root-Nyquist signaling.

While Nyquist signaling achieves excellent spectral efficiencies, there are some implementation-related difficulties associated with it. Since the pulses are not time duration limited, they need to be approximated over some finite time duration, which causes some spectral spillover. A more severe problem occurs with timing errors in the sampling. Inaccurate timing will generate a possibly large number of adjacent symbols to interfere with each other's detection, making timing very crucial. Transmission over nonflat frequency channels also causes more severe intersymbol interference than with some other pulse shapes. This interference needs to be compensated or equalized. Equalization is discussed in standard textbooks on digital communications, e.g., [5, 6].

2.6   THE COMPLEX EQUIVALENT BASEBAND MODEL

In practical applications one often needs to shift a narrowband signal into a higher frequency band for purposes of transmission. The reason for that may lie in the transmission properties of the physical channel, which only allows the passage of signals in certain, usually high, frequency bands. This occurs for example in radio transmission. The process of shifting a signal in frequency is called modulation with a carrier frequency. Modulation is also important for wire-bound transmission, since it allows the coexistence of several signals on the same physical medium, all residing in different frequency bands; this is known as *frequency division multiplexing* (FDM). Probably the most popular modulation method for digital signals is *quadrature double sideband suppressed carrier* (DSB-SC) modulation.

DSB-SC modulation is a simple linear shift in frequency of a signal $x(t)$ with low-frequency content, called a *baseband*, into a higher-frequency band by multiplying $x(t)$ by a cosine or sine waveform with carrier frequency f_0, as shown in Figure 2.7, giving a carrier signal

$$s_0(t) = x(t)\sqrt{2}\cos(2\pi f_0 t), \tag{2.31}$$

where the factor $\sqrt{2}$ is used to make the powers of $s_0(t)$ and $x(t)$ equal.

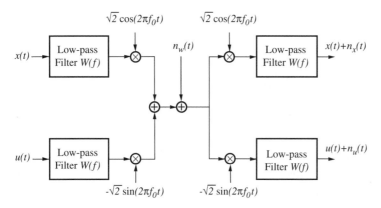

Figure 2.7 Quadrature DSB-SC modulation/demodulation stages.

If our baseband signal $x(t)$ occupies frequencies from 0 to W Hz, $s_0(t)$ occupies frequencies from $f_0 - W$ to $f_0 + W$ Hz, an expansion of the bandwidth by a factor of 2. But we quickly note that we can put another signal, $-u(t)\sqrt{2}\sin(2\pi f_0 t)$, into the same frequency band and that both baseband signals $x(t)$ and $u(t)$ can be recovered by the demodulation operation shown in Figure 2.7, known as a *product demodulator*, where the low-pass filters $W(f)$ serve to reject unwanted out-of-band noise and signals. It can be shown (see, for example, [1]) that this arrangement is optimal; that is, no information or optimality is lost by using the product demodulator for DSB-SC-modulated signals.

If the synchronization between modulator and demodulator is perfect, the signals $x(t)$, the *in-phase* signal, and $u(t)$, the *quadrature* signal, are recovered independently without affecting each other. The DSB-SC-modulated bandpass channel is then, in essence, a dual channel for two independent signals, each of which may carry an independent data stream.

In view of our earlier approach using *basis functions* we may want to view each pair of identical input pulses to the two channels as a two-dimensional signal. Since these two dimensions are intimately linked through the carrier modulation, and since bandpass signals are so ubiquitous in digital communications, a complex notation for bandpass signals has been widely adopted. In this notation, the in-phase signal $x(t)$ is real, and the quadrature signal $ju(t)$ is an imaginary signal, expressed as

$$s_0(t) = x(t)\sqrt{2}\cos(2\pi f_0 t) - u(t)\sqrt{2}\sin(2\pi f_0 t)$$
$$= \mathrm{Re}\left[(x(t) + ju(t))\sqrt{2}e^{2\pi j f_0 t}\right], \tag{2.32}$$

where $s(t) = x(t) + ju(t)$ is called the *complex envelope* of $s_0(t)$.

If both signals are sequences of identical pulses $p(t)$, the complex envelope becomes

$$s(t) = \sum_{r=-l}^{l} (a_r + jb_r)\, p(t - rT) = \sum_{r=-l}^{l} c_r p(t - rT), \qquad (2.33)$$

where c_r is a complex (two-dimensional) number, representing both the in-phase and the quadrature information symbols.

The noise entering the system is demodulated and produces the two low-pass noise waveforms $n_x(t)$ and $n_u(t)$, given by

$$n_x(t) = \sqrt{2} \int_{-\infty}^{\infty} n_w(\alpha) \cos(2\pi f_0 \alpha) w(t - \alpha)\, d\alpha, \qquad (2.34)$$

$$n_u(t) = \sqrt{2} \int_{-\infty}^{\infty} n_w(\alpha) \sin(2\pi f_0 \alpha) w(t - \alpha)\, d\alpha, \qquad (2.35)$$

where $w(t) = \sin(2\pi t W)/2\pi t W$ is the impulse response of an ideal low-pass filter with cutoff frequency W. The autocorrelation function of $n_x(t)$ (and $n_u(t)$ analogously) is given by

$$
\begin{aligned}
\mathrm{E}\,[n_x(t)n_x(t + \tau)] &= 2\int_{-\infty}^{\infty}\int_{-\infty}^{\infty} \cos(2\pi f_0 \alpha)\cos(2\pi f_0 \beta) w(t - \alpha) \\
&\qquad \times w(t + \tau - \beta)\mathrm{E}\,[n_w(\alpha)n_w(\beta)]\, d\alpha\, d\beta \\
&= \frac{N_0}{2} \int_{-\infty}^{\infty} (1 + \cos(4\pi f_0 \alpha)) \\
&\qquad \times w(t - \alpha)w(t + \tau - \alpha)\, d\alpha \\
&= \frac{N_0}{2} \int_{-\infty}^{\infty} w(t - \alpha)w(t + \tau - \alpha)\, d\alpha \\
&= \frac{N_0}{2} \int_{-W}^{W} e^{-2\pi jf\tau}\, df = N_0 W \frac{\sin(2\pi W\tau)}{2\pi W\tau},
\end{aligned} \qquad (2.36)
$$

where we have used Parseval's relationships [1, pp. 237–238] in the third step; that is, the power of $n_x(t)$ equals $\mathrm{E}\,[n^2(t)] = W N_0$. Equation (2.36) is the correlation function of white noise passed through a low-pass filter of bandwidth W; that is, the multiplication with the demodulation carrier has no influence on the statistics of the output noise $n_x(t)$ and can be ignored. Similarly, the cross-correlation function between $n_x(t)$ and $n_u(t)$ is evaluated as

$$
\begin{aligned}
E\left[n_x(t)n_u(t+\tau)\right] &= 2\int_{-\infty}^{\infty}\int_{-\infty}^{\infty}\cos(2\pi f_0\alpha)\sin(2\pi f_0\beta)w(t-\alpha)\\
&\qquad\times w(t+\tau-\beta)E\left[n_w(\alpha)n_w(\beta)\right]d\alpha\,d\beta\\
&= \frac{N_0}{2}\int_{-\infty}^{\infty}\sin(4\pi f_0\alpha)w(t-\alpha)w(t+\tau-\alpha)\,d\alpha\\
&= 0.
\end{aligned}
\tag{2.37}
$$

The system can hence be modeled as two parallel channels disturbed by two independent Gaussian noise processes $n_w^{(x)}(t)$ and $n_w^{(u)}(t)$ as illustrated in Figure 2.8.

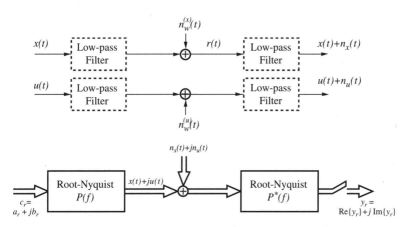

Figure 2.8 Modeling of a DSB-SC system in additive white Gaussian noise by two independent AWGN channels. These two channels can be represented as one complex channel model, shown in the lower part of the figure, and referred to as the equivalent complex baseband model for DSB-SC modulation.

Equivalently we can use a complex model with complex noise $n(t) = n_x(t) + jn_u(t)$, whose correlation is given by

$$
E\left[n(t)n^*(t)\right] = 2N_0W\frac{\sin(2\pi W\tau)}{2\pi W\tau},
\tag{2.38}
$$

which is a shorthand version of (2.36) and (2.37). We thus have finally arrived at the complex equivalent baseband model also shown in Figure 2.8, which takes at its input complex numbers c_r, passes them through complex (dual) modulator and demodulator filters, and feeds the sampled values y_r into a complex receiver. Note that as long as we adopt the convention always to use a receiver filter ($P(f)$ in Figure 2.8), we may omit the low-pass filter in the model, since it can be subsumed into the receiver filter, and the noise source can be made an ideal white Gaussian noise source $(n_x(t) = n_w^{(x)}(t), n_u(t) = n_w^{(u)}(t))$ with correlation function

$$E\left[n(t)n^*(t)\right] = N_0\delta(t). \tag{2.39}$$

According to this discussion, if Nyquist pulses are used with independent complex data symbols c_r, each sample is in fact an independent, two-dimensional signal, hence the ubiquity of two-dimensional signal constellations. Figure 2.9 shows some of the more popular two-dimensional signal constellations. The signal points correspond to sets of possible complex values that c_r can assume.

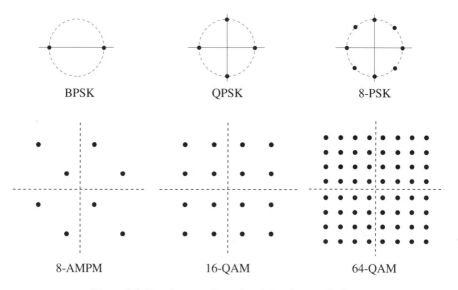

Figure 2.9 Popular two-dimensional signal constellations.

2.7 SPECTRAL BEHAVIOR

Since bandwidth has become an increasingly treasured resource, an important parameter of how efficiently a system uses its allotted bandwidth is the bandwidth efficiency η, defined in (1.2) as

$$\eta = \frac{\text{Bit rate}}{\text{Channel bandwidth } W} \quad \text{[bits/s/Hz].} \tag{2.40}$$

If raised cosine Nyquist signaling is used with a roll-off factor of 0.3, binary phase-shift keying (BPSK) achieves a bandwidth efficiency of $\eta = 0.884$ bits/s/Hz at an E_b/N_0 of 8.4 dB as shown in Figure 1.6. The bandwidth efficiency of QPSK is twice that of BPSK for the same value of E_b/N_0, because

QPSK uses the complex dimension of the signal space. Also, 8-PSK is less power efficient than 8-AMPM (amplitude-modulated phase-modulated) (also called 8-cross) due to the equal energy constraint of the different signals and the resulting smaller Euclidean distances between signal points. Note from Figure 1.6 how bandwidth can be traded for power efficiency, and vice versa, even without applying any coding. All performance points for uncoded signaling lie on a line parallel to the Shannon bound.

To evaluate the spectral efficiency in (2.40), we must first find the power spectrum of the complex pulse train

$$s(t - \alpha) = \sum_{r=-l}^{l} c_r p(t - rT - \alpha), \tag{2.41}$$

where we have introduced a random delay $\alpha \in [0, T[$, and we further assume that the distribution of α is uniform. This will simplify our mathematics and has no influence on the power spectrum, since, surely, knowledge of the delay of $s(t)$ can have no influence on its spectral power distribution.

The advantage of (2.41) is that, if we let $l \to \infty$, $s(t - \alpha)$ becomes a stationary random process with autocorrelation function

$$R_s(t, t + \tau) = \mathrm{E}\left[s^*(t)s(t + \tau)\right] = \mathrm{E}\left[s^*(t - \alpha)s(t + \tau - \alpha)\right]$$

$$= \sum_{p=-\infty}^{\infty} \sum_{q=-\infty}^{\infty} \mathrm{E}\left[c_p^* c_q\right] \mathrm{E}_\alpha\left[p(t - pT - \alpha)p(t + \tau - qT - \alpha)\right]. \tag{2.42}$$

Let us assume that the discrete autocorrelation of the symbols c_r is stationary; that is,

$$R_{cc}(r) = \mathrm{E}\left[c_q^* c_{q+r}\right] \tag{2.43}$$

depends only on r. This lets us rewrite (2.42) as

$$R_s(t, t + \tau) = \sum_{r=-\infty}^{\infty} R_{cc}(r) \sum_{q=-\infty}^{\infty} \mathrm{E}_\alpha\left[p(t - qT - \alpha)p(t + \tau - (q + r)T - \alpha)\right]$$

$$= \sum_{r=-\infty}^{\infty} R_{cc}(r) \sum_{q=-\infty}^{\infty} \frac{1}{T} \int_0^T p(t - qT - \alpha)p(t + \tau - (q+r)T - \alpha)\, d\alpha$$

$$= \sum_{r=-\infty}^{\infty} R_{cc}(r) \frac{1}{T} \int_0^T p(\alpha)p(\alpha + \tau - rT)\, d\alpha, \tag{2.44}$$

where the latter term, $\int_0^T p(\alpha)p(\alpha + \tau - rT)\, d\alpha = R_{pp}(\tau - rT)$, is the pulse autocorrelation function of $p(t)$, which depends only on τ and r, making $R_s(t, t + \tau) = R_s(\tau)$ stationary.

To find the power spectral density of $s(t)$, we now merely need the Fourier transform of $R_s(\tau)$; that is,

$$\Phi_{cc}(f) = \frac{1}{T} \int_{-\infty}^{\infty} \sum_{r=-\infty}^{\infty} R_{cc}(r) R_{pp}(\tau - rT) e^{-j2\pi f\tau} \, d\tau$$

$$= \frac{1}{T} \sum_{r=-\infty}^{\infty} R_{cc}(r) e^{-j2\pi frT} \int_{-\infty}^{\infty} R_{pp}(\tau) e^{-j2\pi f\tau} \, d\tau$$

$$= \frac{1}{T} C(f) G(f),$$

where

$$C(f) = \sum_{r=-\infty}^{\infty} R_{cc}(r) e^{-j2\pi frT} \qquad (2.45)$$

is the spectrum-shaping component resulting from the correlation of the complex symbol sequences c_r, and $G(f) = |P(f)|^2$ is the energy spectrum of the symbol pulse $p(t)$ (page 31).

While the spectral factor due to the correlation of the pulses, $C(f)$, can be used to help shape the signal spectrum as in partial-response signaling [5, pp. 548ff.] or correlative coding [11], $R_{cc}(r)$ is often an impulse δ_{0r}, corresponding to choosing uncorrelated, zero-mean-valued symbols c_r, and $C(f)$ is flat. If this is the case, the spectrum of the transmitted signal is exclusively shaped by the symbol pulse $p(t)$. This holds, for example, for QPSK if the symbols are chosen independently and with equal probability. In general it holds for any constellation and symbol probabilities whose discrete autocorrelation $R_{cc}(r) = \delta_{0r}$ and whose symbol mean[2] $E[c_q] = 0$. Constellations for which this is achieved by the uniform probability distribution for the symbols are called *symmetric constellations* (e.g., M-PSK, 16-QAM (quadrature amplitude modulation), 64-QAM, etc.). We will see in Chapter 3 that the uncorrelated nature of the complex symbols c_r is the basis for the fact that trellis-coded modulation does not shape the signal spectrum (also discussed in [12]).

APPENDIX

The random waveform $n(t)$ that is added to the transmitted signal in the channel of Figure 2.1 can be modeled as a Gaussian random process in

---

[2] If the mean of the symbols c_r is not equal to zero, discrete frequency components appear at multiples of $1/T$ (see [5, Section 3.4]).

many important cases. Let us start by approximating this noise waveform by a sequence of random rectangular pulses $w(t)$ of duration T_s; that is,

$$n(t) = \sum_{i=-\infty}^{\infty} n_i w(t - iT_s - \delta), \tag{2.46}$$

where the weighing coefficients n_i are zero-mean, independent Gaussian random variables, T_s is the chosen discrete-time increment, and δ is some random delay, uniformly distributed in $[0, T_s[$. The random delay δ is needed to make (2.46) stationary. A sample function of $n(t)$ is illustrated in Figure 2.10.

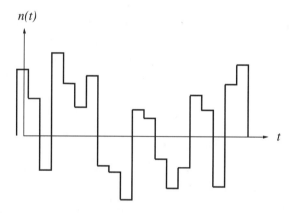

Figure 2.10 Sample function of the discrete approximation to the additive channel noise $n(t)$.

A stationary Gaussian process is completely determined by its mean and correlation function (see, e.g., [2, 3]). The mean of the noise process is assumed to be zero, and the correlation function can be calculated as

$$R(\tau) = \mathrm{E} \left[\sum_{i=-\infty}^{\infty} \sum_{j=-\infty}^{\infty} n_i n_j w(t - iT_s - \delta) w(t + \tau - jT_s - \delta) \right]$$

$$= \sigma^2 \mathrm{E}_\delta \left[\sum_{i=-\infty}^{\infty} w(t - iT_s - \delta) w(t + \tau - iT_s - \delta) \right] \tag{2.47}$$

$$= \sigma^2 \Delta_{T_s}(\tau),$$

where $\sigma^2 = \mathrm{E}\left[n_i^2\right]$ is the variance of the discrete noise samples n_i and the expectation is now only over the variable delay δ. This expectation is easily evaluated, and $\Delta_{T_s}(\tau)$ is a triangular function as shown in Figure 2.11.

We note that the power of the noise function, $P = R(\tau = 0) = \sigma^2/T_s$, increases linearly with the inverse of the sample time T_s. To see why that is a reasonable effect, let us consider the power spectral density $N(f)$ of the random process $n(t)$, given by the Fourier transform of the correlation function $R(\tau)$ as

$$N(f) = \int_{-\infty}^{\infty} R(\tau)e^{-2\pi j\tau f}\, d\tau, \tag{2.48}$$

where f is the frequency variable. The power density spectrum $N(f)$ is also shown in Figure 2.11. Note that as $T_s \to 0$ in the limit, $N(f) \to \sigma^2$. In the limit approximation, (2.46) therefore models noise with an even distribution of power over all frequencies. Such noise is called *white noise* in analogy to the even-frequency content of white light, and we will denote it by $n_w(t)$ henceforth. In the literature the constant $\sigma^2 = N_0/2$, and N_0 is known as the one-sided noise power spectral density [1].

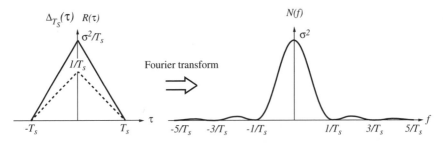

Figure 2.11 Correlation function $R(\tau)$ and power spectral density $N(f)$ of the discrete approximation to the noise function $n(t)$.

Naturally, the model makes no physical sense in the limit, since it would imply infinite noise power. But we will be careful not to use the white noise $n_w(t)$ without some form of low-pass filtering. If $n_w(t)$ is filtered, the high noise frequencies are rejected in the stopbands of the filter, and it is irrelevant for the filter output how we model the noise in its stopbands.

As $n(t) \to n_w(t)$, the correlation function $R(\tau)$ will degenerate into a pulse of width zero and infinite height as $T_s \to 0$; i.e., $R(\tau) \to \sigma^2\delta(\tau)$, where $\delta(\tau)$ is known as *Dirac's impulse function*. $\delta(\tau)$ is technically not a function but a distribution. We will only need the *sifting property* of $\delta(t)$; that is,

$$\int_{-\infty}^{\infty} \delta(t - \alpha)f(t)\, dt = f(\alpha), \tag{2.49}$$

where $f(t)$ is an arbitrary function, which is continuous at $t = \alpha$. Property (2.49) is easily proven by using (2.47) and carrying out the limit operation in the integral (2.49). In fact, the relation

$$\int_a^b \delta(\tau - \alpha) f(\tau)\, d\tau = \lim_{T_s \to 0} \int_a^b \Delta_{T_s}(\tau - \alpha) f(\tau)\, d\tau = f(\alpha), \qquad (2.50)$$

for any α in the interval (a, b), can be used as a proper definition of the impulse function. If the limit in (2.50) could be taken inside the integral, then

$$\delta(\tau - \alpha) = \lim_{T_s \to 0} \Delta_{T_s}(\tau - \alpha). \qquad (2.51)$$

However, any function which is zero every where except at one point equals zero when integrated in the Riemann sense, and, hence, equation (2.51) is a symbolic equality only, to be understood in the sense of (2.50). An introductory discussion of distributions can be found, for example, in [4, pp. 269–282].

References

[1] J. M. Wozencraft and I. M. Jacobs, *Principles of Communication Engineering*, Wiley, New York, 1965; reprinted by Waveland Press, 1993.

[2] W. Feller, *An Introduction to Probability Theory and Its Applications*, Vols. 1, 2, rev. printing of 3rd ed., Wiley, New York, 1970.

[3] K. S. Shanmugan and A. M. Breipohl, *Random Signals: Detection, Estimation and Data Analysis*, Wiley, New York, 1988.

[4] A. Papoulis, *The Fourier Integral and Its Applications*, McGraw-Hill, New York, 1962.

[5] J. G. Proakis, *Digital Communications*, 3rd ed., McGraw-Hill, New York, 1995.

[6] B. Sklar, *Digital Communications: Fundamentals and Applications*, Prentice Hall, Englewood Cliffs, NJ, 1988.

[7] R. G. Gallager, *Information Theory and Reliable Communication*, Wiley, New York, 1968.

[8] C. E. Shannon, "A mathematical theory of communications," *Bell Syst. Tech. J.*, Vol. 27, pp. 379–423, 1948.

[9] R. E. Blahut, *Digital Transmission of Information*, Addison-Wesley, New York, 1990.

[10] S. Ramseier and C. Schlegel, "On the bandwidth/power tradeoff of trellis coded modulation schemes," *Proc. IEEE Globecom'93*, 1993.

[11] S. Ramseier, "Bandwidth-efficient correlative trellis coded modulation schemes," *IEEE Trans. Commun.*, Vol. COM-42, No. 2/3/4, pp. 1595–1605, 1994.

[12] E. Biglieri, "Ungerböck codes do not shape the signal power spectrum," *IEEE Trans. Inform. Theory*, Vol. IT-32, No. 4, pp. 595–596, 1986.

CHAPTER 3

TRELLIS-CODED MODULATION

3.1 AN INTRODUCTORY EXAMPLE

As we have seen in Chapter 1, power and bandwidth are limited resources in modern communications systems, and efficient exploitation of these resources will invariably involve an increase in the complexity of a communication system. It has become apparent over the past few decades that while there are strict limits on the power and bandwidth resources, the complexity of systems could steadily be increased to obtain efficiencies ever closer to the theoretical limits. One very successful method of reducing the power requirements without increase in the requirements on bandwidth was introduced by Gottfried Ungerböck [1–4] and was subsequently termed *trellis-coded modulation* (TCM). We start this chapter with an illustrative example of this intriguing method of combining coding and modulation.

Let us assume that we are using a standard quadrature phase-shift keying (QPSK) modulation scheme, which allows us to transmit two information bits per symbol. We know from elementary communication theory that the most likely error a decoder makes due to noise is to confuse two neighboring signals, say signal $\mathbf{s}^{(1)}$ and signal $\mathbf{s}^{(2)}$ (compare Figure 2.2). This will happen with probability

$$P_{\mathbf{s}^{(1)} \to \mathbf{s}^{(2)}} = Q\left(\sqrt{\frac{d^2 E_s}{2N_0}}\right), \tag{3.1}$$

where $d^2 = 2$ for a unit energy signal constellation,[1] and E_s is the average energy per symbol. Instead of using such a system by transmitting one symbol at a time, the encoder shown in Figure 3.1 is used. This encoder consists of two parts, the first of which is a *finite-state machine* (FSM) with a total of eight states, where state s_r at time r of the FSM is defined by the contents of the delay cells; that is, $s_r = (s_r^{(2)}, s_r^{(1)}, s_r^{(0)})$. The second part is called a *signal mapper*, and its function is a memoryless mapping of the 3 bits $v_r = (u_r^{(2)}, u_r^{(1)}, v_r^{(0)})$ into one of the eight symbols of an 8-PSK signal set. The FSM accepts 2 input bits $u_r = (u_r^{(2)}, u_r^{(1)})$ at each symbol time r and transits from a state s_r to one of four possible successor states s_{r+1}. In this fashion the encoder generates the (possibly) infinite sequence of symbols $\mathbf{x} = (\ldots, x_{-1}, x_0, x_1, x_2, \ldots)$. Assuming that the encoder is operated in a continuous fashion, there are four choices at each time r, which allows us to transmit two information bits/symbol, the same as with QPSK.

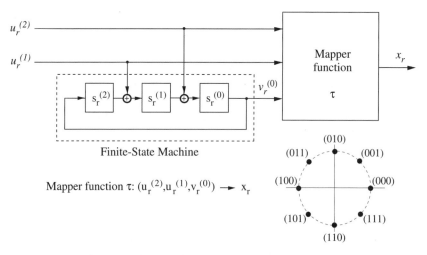

Figure 3.1 Trellis encoder with an 8-state finite-state machine driving a 3-bit to 8-PSK signal mapper.

A graphical interpretation of the operation of this FSM, and in fact the entire encoder, will prove immensely useful. Since the FSM is time invariant, it can be represented by a state-transition diagram as shown in Figure 3.2. The nodes in this transition diagram are the states of the FSM, and the branches represent the possible transitions between states. Each branch can now be labeled by the pair of input bits $u = (u^{(2)}, u^{(1)})$ that

---

[1] We now assume that all signal constellations are normalized, so that their average energy is unity, and $\sqrt{E_s}$ is the amplification factor that generates the average signal energy E_s.

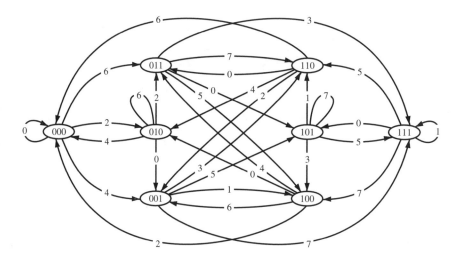

Figure 3.2 State-transition diagram of the encoder from Figure 3.1. The labels on the branches are the encoder output signals $x(v)$ in decimal notation.

cause the transition and by either the output triple $v = (u^{(2)}, u^{(1)}, v^{(0)})$ or the output signal $x(v)$. (In Figure 3.2 we have used $x(v)$, represented in decimal notation; i.e., $x_{\text{dec}}(v) = u^{(2)}2^2 + u^{(1)}2^1 + v^{(0)}2^0$.)

If we index the states by their content and the time index r, Figure 3.2 expands into the *trellis diagram*, or simply the *trellis* of Figure 3.3. The trellis of a trellis encoder is therefore the two-dimensional representation of the operation of the encoder, it captures all possible state transitions starting from an originating state (usually state 0) and terminating in a final state (usually also state 0). The trellis in Figure 3.3 is terminated to length $L = 5$. This requires that the FSM be driven back into the state 0 at time $r = 5$. As a consequence, the branches at times $r = 3$ and 4 are predetermined, and no information is transmitted in those two time units. Contrary to the short trellis in Figure 3.3, in practice the length of the trellis will be several hundred or thousands of time units, possibly even infinite, corresponding to continuous operation. When and where to terminate the trellis are practical considerations.

Now each path through the trellis corresponds to a unique message sequence and is associated with a unique signal sequence (compare Section 2.5). The term *trellis-coded modulation* originates from the fact that these encoded sequences consist of modulated symbols rather than binary digits.

At first sight it is not clear what has been gained by doing this compli-cated encoding since the signals in the 8-PSK signal sets are much closer than those in the QPSK signal set and have a higher symbol error rate. The

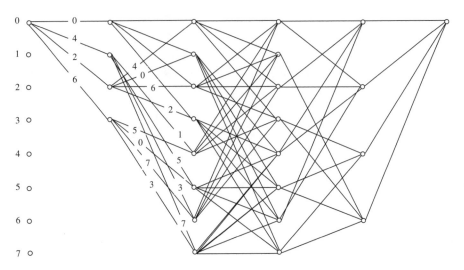

Figure 3.3 Trellis diagram of the encoder from Figure 3.1. The code is terminated
to length $L = 5$.

FSM, however, puts restrictions on the symbols that can be in a sequence,
and these restrictions can be exploited by a smart decoder. In fact, what
counts is the distance between signal sequences **x**, not the distance between
individual signals. Let us then assume that such a decoder can follow all
possible sequences through the trellis and that it makes decisions between
sequences. This is illustrated in Figure 3.4 for sequences $\mathbf{x}^{(e)}$ (erroneous)
and $\mathbf{x}^{(c)}$ (correct). These sequences differ in the three symbols shown.
An optimal decoder will make an error between these two sequences with
probability $P_s = Q\left(\sqrt{d_{ec}^2 E_s/2N_0}\right)$, where $d_{ec}^2 = 4.586 = 2 + 0.56 + 2$
is the Euclidean distance between $\mathbf{x}^{(e)}$ and $\mathbf{x}^{(c)}$, which, incidently, is much
larger than the QPSK distance of $d^2 = 2$. Going through all possible se-
quence pairs $\mathbf{x}^{(e)}$ and $\mathbf{x}^{(c)}$, one finds that those highlighted in Figure 3.4
have the smallest squared Euclidean distance, and, hence, the probability
that the decoder makes an error between those two sequences is the most
likely error.

    We now see that by virtue of doing sequence decoding rather than
symbol decoding the distances between alternatives can be increased, even
though the signal constellation used for sequence coding has a smaller
minimum distance between signal points. For this code we may decrease
the symbol power by about 3.6 dB; that is, we use less than half the power
needed for QPSK.

    A more precise error analysis is not quite so trivial since the possible
error paths in the trellis are highly correlated, which makes an exact analysis

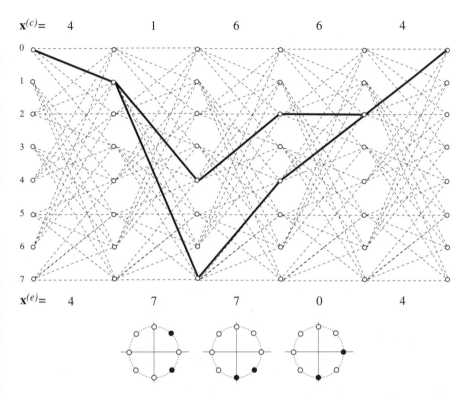

Figure 3.4 Section of the trellis of the encoder in Figure 3.1. The two solid
lines depict two possible paths with their associated signal sequences
through this trellis. The numbers on top are the signals transmitted
if the encoder follows the upper path, and the numbers at the bottom
are those on the lower path.

of the error probability impossible for all but the simplest cases. In fact, a
lot of work has gone into analyzing the error behavior of trellis codes, and
we devote an entire chapter (Chapter 5) to this topic. This simple example
should convince us, however, that some real coding gain can be achieved
by this method with a relatively manageable effort. Having said that, we
must stress that the lion's share of the work is performed by the decoder,
about which we will say more later.

3.2   GROUP-TRELLIS CODES

While there exist an infinite number of ways to implement the FSM that
introduces the symbol correlation, we mainly concentrate on the subclass
of FSMs that generate a group trellis (defined later). This restriction sounds
quite severe, but we do this for two reasons. First, by far the largest part

of the published literature over the past decade deals with trellis codes of this kind; second, we show in Chapter 5 that there exist group-trellis codes that can achieve the capacity of an additive white Gaussian noise channel using a time-varying mapping function. There are important examples of trellis codes that do not use a group trellis, such as the nonlinear rotationally invariant codes (discussed in Section 3.7).

Let us then start by considering a group trellis. The group[2] property of the class of trellises considered here relates to their topology in the sense that such a trellis "looks" the same for every possible path through the trellis. It is symmetrical in that every path has the same number of neighbor paths of a given length. This concept is most easily captured by assigning different elements from a group S to different states and assigning elements (not necessarily different) from a group B to the branches. A trellis is then a group trellis if the paths through the trellis, described by sequences of pairs $(b_0, s_0), (b_1, s_1), \ldots, (b_L, s_L)$, also form a group, where the group operation is the elementwise group operation in the component groups, denoted by $\oplus$. This then gives us the following definition.

DEFINITION 3.1

The trellis generated by a finite-state machine is a group trellis if, for any L, there exists a labeling of paths $p^{(i)} = (b_0, s_0), \ldots, (b_L, s_L)$, where $b_r \in B$ and $s_r \in S$ such that $e = p^{(i)} \oplus p^{(j)} = (b_0^{(1)} \oplus b_0^{(2)}, s_0^{(1)} \oplus s_0^{(2)}), \ldots, (b_L^{(1)} \oplus b_L^{(2)}, s_L^{(1)} \oplus s_L^{(2)})$ is also a valid path in this trellis. The $\oplus$-operation is taken as the appropriate group operation in B or S, respectively.

Note that this definition could be simplified by considering only the sequences of states that also uniquely determine a path through the trellis. The problem with such a definition is that the mapper function can no longer be made memoryless, since it would depend on the previous state. In any case, the labeling $p^{(i)} = (s_1, s_0), (s_2, s_1), (s_3, s_2), \ldots$ is perfectly valid.

As an example consider the trellis generated by the FSM of the previous section as illustrated again in Figure 3.5. The state label is the decimal representation of the binary content of the three delay cells, e.g., $5 \equiv (101)$,

---

[2] A group G is a set of elements with the following properties:

(i)   Closure under a binary group operation (usually called addition, and denoted by $\oplus$); i.e., if $g_i, g_j \in G$, then $g_i \oplus g_j = g_l \in G$.

(ii) There exists unit element 0 such that $g \oplus 0 = g$.

(iii) Every element g possesses an additive inverse $(-g) \in G$, such that $g \oplus (-g) = 0$.

(iv) (For Abelian groups) The addition operation commutes: $g_i \oplus g_j = g_j \oplus g_i$.

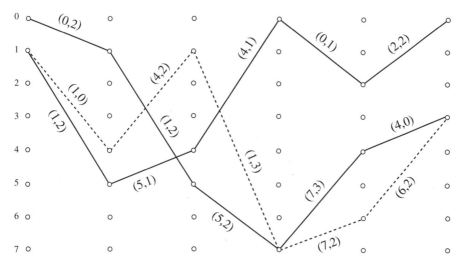

Figure 3.5 Additive path labeling in a group trellis.

and addition is the bitwise XOR operation (GF(2) addition). The branch labels b are likewise the decimal representation of the two binary input digits; that is, $b = u$. Adding the two solid paths in Figure 3.5 results in the dotted path, which is also a valid path.

All we need is some FSM to generate a group trellis, and that can conveniently be done by a structure such as the one in Figure 3.6. In general, a group-trellis encoder can be built as shown in Figure 3.6a for an FSM with 2^v states and 2 input bits/symbol, where the mapper function is an arbitrary memoryless function $x_r = \tau(u_r, s_r)$ of the current state of the FSM and the new input digits $u_r = (u_r^{(k)}, \ldots, u_r^{(1)})$. Figure 3.6b shows how trellis codes were originally constructed by using only the output bits $(u_r, s_r) = v_r = (v_r^{(n)}, v_r^{(n-1)}, \ldots, v_r^{(0)})$ in the mapper function $x_r = \tau(v_r)$. This is clearly a subclass of the class of general encoders. The reason why the construction of trellis codes was approached in this fashion was that the FSM implementation in 3.6b is in effect a convolutional encoder.[3] Convolutional codes have enjoyed tremendous success in coding for noisy channels. Their popularity has made them the subject of numerous papers and books, and we will discuss convolutional codes in more detail in Chapter 4. For our purpose, however, they mainly serve as generators of the group trellis; that is, if a

_____

[3] More precisely, the generating FSM is a convolutional encoder in the systematic feedback form (compare Chapter 4).

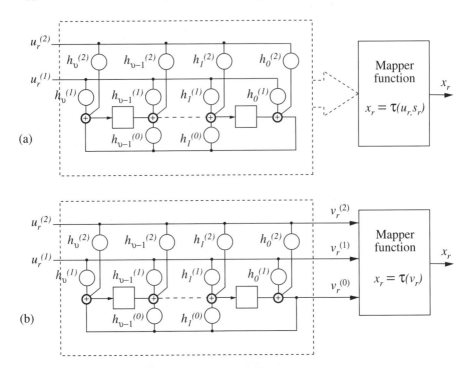

Figure 3.6 General structure of a group-trellis encoder and the structure originally used to construct trellis codes.

convolutional code is used to generate the trellis, it will always be a group trellis.

3.3   THE MAPPER FUNCTION

So far we have not said much about the mapper function $x_r = \tau(u_r, s_r)$ that assigns a signal x_r to each transition of the FSM. The mapper function of the trellis code discussed in the beginning of this chapter maps 3 output bits from the FSM into one 8-PSK signal point as shown in Figure 3.7.

The important quantities of a signal set are the squared Euclidean distances between pairs of signal points, since they determine the squared Euclidean distances between sequences, which are the important parameters of a trellis code.

A signal on a branch in the trellis $x^{(i)} = \tau(b^{(i)}, s^{(i)})$ is generated by the mapper function τ, given as input the branch label $\tau(b^{(i)}, s^{(i)})$. We now wish to explore how the distance between two branch signals $d^2 = \left\| x^{(1)} - x^{(2)} \right\|^2$ is related to the branch labels. Now, if the squared

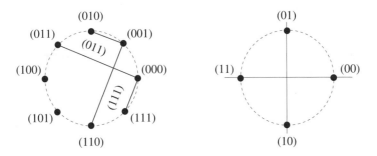

Figure 3.7 8-PSK and QPSK signal sets with natural and Gray mappings.

Euclidean distance between $x^{(1)}$ and $x^{(2)}$, given by

$$d^2 = \left\| \tau(b^{(1)}, s^{(1)}) - \tau(b^{(2)}, s^{(2)}) \right\|^2, \qquad (3.2)$$

depends only on the difference between the branch labels—that is,

$$
\begin{aligned}
d^2 &= \left\| \tau(b^{(1)}, s^{(1)}) - \tau(b^{(2)}, s^{(2)}) \right\|^2 \\
&= \left\| \tau(b^{(1)} \oplus (-b^{(2)}), s^{(1)} \oplus (-s^{(2)})) - \tau(0,0) \right\|^2 \\
&= f(e_b, e_s), \qquad e_b = b^{(1)} \oplus (-b^{(2)}), e_s = s^{(1)} \oplus (-s^{(2)}) \qquad (3.3)
\end{aligned}
$$

—then the signal $\tau(0, 0)$ can always be used as a reference signal. Signal mappings with this property are called *regular*. An example of a regular mapping is the QPSK signal set in Figure 3.7 with Gray mapping. It is easy to see that a binary difference of (01) and (10) will always produce a $d^2 = 2$. The binary difference of (11) produces $d^2 = 4$. The 8-PSK signal set in Figure 3.7 uses a mapping known as *natural mapping*, where the signal points are numbered sequentially in a counterclockwise fashion. Note that this mapping is not regular, since the binary difference $v = (011)$ can produce the two distances $d^2 = 0.56$ and $d^2 = 3.14$, depending on $(b^{(1)}, s^{(1)})$. We will see (Section 3.8) that no regular mapping exists for an 8-PSK signal set.

If the labels of a group trellis are used to drive a regular mapper, we obtain a *regular* trellis code, which has the property that the set of distances from a given path $p^{(i)}$ to all other paths does not depend on $p^{(i)}$ by virtue of the additivity of the branch labels that are mapped into a regular signal set. In fact, such a code is a generalized group code,[4] also known as a

---

[4] A code $\mathcal{C}$ is a group code if there exists a group operation $\oplus$, such that the codewords in $\mathcal{C}$ form a group. A regular code is a generalized group code in the following sense: Even though the additivity among codewords (sequences) $\mathbf{x}^{(i)}$ does not hold, the weaker, generalized relation $d^2(\mathbf{x}(\mathbf{v}^{(i)}), \mathbf{x}(\mathbf{v}^{(j)})) = d^2(\mathbf{x}(\mathbf{v}^{(i)} \oplus -\mathbf{v}^{(j)}), \mathbf{x}(\mathbf{0}))$, concerning only squared Euclidean distances holds.

geometrically uniform code (Section 3.8). This makes code searches and error performance analysis much easier since we now can concentrate on one arbitrary correct path and know that the distances of possible error paths to it are independent of the actual path chosen. In practice, one usually chooses the path whose labels are the all-zero pairs, that is, the all-zero sequence $(0, 0), \ldots, (0, 0)$.

This simplification is not possible for 8-PSK due to the nonregularity of the mapping. One has an intuitive feeling though that the nonregularity of the 8-PSK signal set is not severe. Let us look again at equation (3.3), and instead of looking at individual distances between pairs we look at the set of distances as we vary over signal pairs on the branches $(b^{(2)}, s^{(2)})$ and $b^{(2)} \oplus e_b, s^{(1)}$ by varying through all $b^{(2)}$; that is, we look at the set

$$
\begin{aligned}
\{d^2\} &= \left\{ \left\| \tau(b^{(2)}, s^{(2)}) - \tau(b^{(2)} \oplus e_b, s^{(1)}) \right\|^2 \right\} \\
&= \left\{ \left\| \tau(v^{(2)}) - \tau(v^{(2)} \oplus e_v) \right\|^2 \right\},
\end{aligned} \tag{3.4}
$$

where $v^{(i)} = (b^{(i)}, s^{(i)})$ and $e_v = (e_b, e_s)$. Doing this, for example, for $e_v = (011)$, we obtain the set $\{0.56, 0.56, 3.14, 3.14\}$, which is independent of the state $s^{(2)}$ and depends only on the branch label difference e_v. This can be done for all label differences e_v, and we find

e_v	$\{d^2\}$
000	$\{0,0,0,0\}$
001	$\{0.56,0.56,0.56,0.56\}$
010	$\{2,2,2,2\}$
011	$\{0.56,0.56,3.14,3.14\}$
100	$\{4,4,4,4\}$
101	$\{3.14,3.14,3.14,3.14\}$
110	$\{2\}$
111	$\{0.56,0.56,3.14,3.14\}$

The different sets[5] of distances thus obtained are all independent of the actual states $s^{(1)}$ and $s^{(2)}$ and depend only on e_v. As we will see in Chapter 5, this is a useful generalization of the notion of regularity, and signal sets with this property are called *quasi-regular*.

DEFINITION 3.2

A signal set is called quasi-regular if all sets of distances $\{d^2_{e_b,e_s}\} = \{\|\tau(u, s_i) - \tau(u \oplus e_b, s_i \oplus e_s)\|^2\}$ between the signals $\tau(u, s_i)$ and $\tau(u \oplus$

---

[5] Strictly speaking $\{0.56, 0.56, 3.14, 3.14\} = \{2 \times 0.56, 2 \times 3.14\}$ is called a family since we also keep multiple occurrences of each entry, but for simplification we will refer to it as a set.

e_b, $s_i \oplus e_s$) from the states s_i and $s_i \oplus e_s$, as we vary over all pairs of branches (u, s_i) and $(u \oplus e_b, s_i \oplus e_s)$, are independent of s_i and depend only on the label difference e_b, e_s.

This definition may not sound very useful at the moment, but we will see in Chapter 5 that, since we have to average over all inputs u in order to obtain an average error performance, the independence of the sets of distances from the correct state s_i, together with the additivity of the group trellis, will allow us to substantially reduce the complexity of distance search algorithms and error performance calculations.

Having defined quasi-regular signal sets we would like to know which signal sets are quasi-regular. This is, in general, not a simple task since we might have to relabel the signal points. But a large class of commonly used signal sets are quasi-regular, for example:

THEOREM 3.1
An MPSK signal set with natural mapping is quasi-regular for $M = 2^n$.

Proof
We have already shown via example that 8-PSK with natural mapping is quasi-regular. That this is the case in general can easily be seen by inspection.

3.4   CONSTRUCTION OF CODES

From Figure 3.4 we see that an error path diverges from the correct path at some state and merges with the correct path again at a (possibly) different state. The task of designing a good trellis code now crystallizes into designing a trellis code for which different symbol sequences are separated by large squared Euclidean distances. Of particular importance is the minimum squared Euclidean distance, termed d_{free}^2; that is, $d_{\text{free}}^2 = \min_{\mathbf{x}^{(i)}, \mathbf{x}^{(j)}} \|\mathbf{x}^{(i)} - \mathbf{x}^{(j)}\|^2$. A code with a large d_{free}^2 is generally expected to perform well, and d_{free}^2 has become the major design criterion for trellis codes. (The minimum squared Euclidean distance for the 8-PSK trellis code of Section 3.1 is $d_{\text{free}}^2 = 4.586$.)

One heuristic design rule [2], which was used successfully in designing codes with large d_{free}^2, is based on the following observation: If we assign to the branches leaving a state signals from a subset where the distances between points are large, and likewise assign such signals to the branches merging into a state, we are assured that the total distance is at least the sum of the minimum distances between the signals in these subsets. For our 8-PSK code example we can choose these subsets to be QPSK signal subsets of the original 8-PSK signal set. This is done by partitioning the 8-PSK signal set into two QPSK sets, as illustrated in Figure 3.8. The mapper function is now chosen such that the state information bit $v^{(0)}$ selects the subset and the input bits u select a signal within the subset. Since all branches leaving a state have the same state information bit $v^{(0)}$, all the branch signals are either in subset A or subset B, and the difference between two signal sequences picks up an incremental distance of $d^2 = 2$ over the first branch of their difference. To achieve this, we need to make sure that u does not affect $v^{(0)}$, which is done by setting $h_0^{(2)} = 0$ and $h_0^{(1)} = 0$ in Figure 3.6.

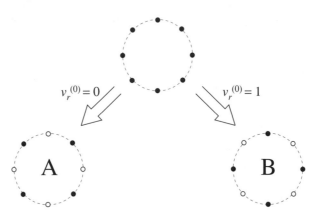

Figure 3.8 An 8-PSK signal set partitioned into constituent QPSK signal sets.

To guarantee that the signal on branches merging into a state can also be chosen from one of these two subsets, we set $h_\nu^{(2)} = 0$ and $h_\nu^{(1)} = 0$. This again has the effect that merging branches have the same value of the state information bit $v^{(0)}$. These are Ungerböck's [2] original design rules.

We have now ensured that the minimum distance between any two paths is at least twice that of the original QPSK signal set. The values of the remaining connector coefficients $h^{(2)} = h_1^{(2)}, \ldots, h_{\nu-1}^{(2)}$, $h^{(1)} = h_1^{(1)}, \ldots, h_{\nu-1}^{(1)}$, and $h^{(0)} = h_1^{(0)}, \ldots, h_{\nu-1}^{(0)}$ are much harder to find, and one usually resorts to computer search programs or heuristic construction algorithms. Table 3.1 shows the best 8-PSK trellis codes found to date using 8-PSK with natural mapping. The figure gives the connector coefficients, d_{free}^2, the average path multiplicity $A_{d_{\text{free}}}$ of d_{free}^2, and the average bit multiplicity $B_{d_{\text{free}}}$ of d_{free}^2. $A_{d_{\text{free}}}$ is the average number of paths at distance d_{free}^2, and $B_{d_{\text{free}}}$ is the average number of bit errors on those paths. Both $A_{d_{\text{free}}}$ and $B_{d_{\text{free}}}$, as well as the higher-order multiplicities, are important

TABLE 3.1   Connectors, Free Squared Euclidean Distance, and Asymptotic Coding Gains of Some Maximum Free-Distance 8-PSK Trellis Codes

Number of states	$h^{(0)}$	$h^{(1)}$	$h^{(2)}$	d^2_{free}	$A_{d_{free}}$	$B_{d_{free}}$	Asymptotic coding gain (dB)
4	5	2	—	4.00^*	1	1	3.0
8	11	2	4	4.59^*	2	7	3.6
16	23	4	16	5.17^*	2.25	11.5	4.1
32	45	16	34	5.76^*	4	22.25	4.6
64	103	30	66	6.34^*	5.25	31.125	5.0
128	277	54	122	6.59^*	0.5	2.5	5.2
256	435	72	130	7.52^*	1.5	12.25	5.8
512	1525	462	360	7.52^*	0.313	2.75	5.8
1024	2701	1216	574	8.10^*	1.32	10.563	6.1
2048	4041	1212	330	8.34	3.875	21.25	6.2
4096	15201	6306	4112	8.68	1.406	11.758	6.4
8192	20201	12746	304	8.68	0.617	2.711	6.4
32768	143373	70002	47674	9.51	0.25	2.5	6.8
131072	616273	340602	237374	9.85			6.9

* The codes with an asterisk were found by exhaustive computer searches [4, 8]; the other codes were found by various heuristic search and construction methods [8, 9]. The connector polynomials are in octal notation.

parameters determining the error performance of a trellis code. This is discussed in detail in Chapter 5. (The connector coefficients are given in octal notation; e.g., $h^{(0)} = 23 = 10111$, where a 1 means connected and a 0 means disconnected.)

From Table 3.1 one can see that an asymptotic coding gain (coding gain for SNR $\rightarrow \infty$ over the reference constellation used for uncoded transmission at the same rate) of about 6 dB can quickly be achieved with moderate effort. But for optimal decoding the complexity of the decoder grows proportionally with the number of states, so it becomes very hard to go significantly beyond 6 dB. Since the asymptotic coding gain is a reasonable yardstick at the bit error rates of interest, codes with a maximum of about 1000 states seem to exploit most of what can be gained by this type of coding.

Other researchers have used different mapper functions to improve performance, particularly bit error performance, which could be improved by up to 0.5 dB. The 8-PSK Gray mapping (label the 8-PSK symbols successively by (000), (001), (011), (010), (110), (111), (101), (100)) was used by Du and Kasahara [5] and Zhang [6, 7]. Zhang also used another mapper (label the symbols successively by (000), (001), (010), (011), (110), (111), (100), (101)) to further improve bit multiplicity. The search criterion employed involved minimizing the bit multiplicities of several spectral lines in the distance spectrum of a code.[6] Table 3.2 gives the best 8-PSK codes found with respect to bit error probability.

TABLE 3.2   Table of Improved 8-PSK Codes Using a Different Mapping Function [6]

Number of states	$h^{(0)}$	$h^{(1)}$	$h^{(2)}$	d^2_{free}	$A_{d_{\text{free}}}$	$B_{d_{\text{free}}}$
8	17	2	6	4.59^*	2	5
16	27	4	12	5.17^*	2.25	7.5
32	43	4	24	5.76^*	2.375	7.375
64	147	12	66	6.34^*	3.25	14.8755
128	277	54	176	6.59^*	0.5	2
256	435	72	142	7.52^*	1.5	7.813
512	1377	304	350	7.52^*	0.0313	0.25
1024	2077	630	1132	8.10^*	0.2813	1.688

If we go to higher-order signal sets, such as 16-quadrature-amplitude modulation, 32-cross, 64-QAM etc., there are, at some point, not enough states left such that each diverging branch leads to a different state, and we have parallel transitions, that is, two or more branches connecting two states. Naturally we would want to assign signals with large distances to such parallel branches to avoid a high probability of error, since the probability of these errors cannot be influenced by the code.

The situation of parallel transition is actually the case for the first 8-PSK code in Table 3.1, whose trellis is given in Figure 3.9. Here the parallel transitions are by choice, though not by necessity. Note that the minimum-distance path pair through the trellis has $d^2 = 4.56$, but that is not the most likely error to happen. All signals on parallel branches are from a BPSK subset of the original 8-PSK set, and hence their distance is $d^2 = 4$, which gives the 3-dB asymptotic coding gain of the code over QPSK ($d^2 = 2$).

In general, we partition a signal set into a partition chain of subsets, such that the minimum distance between signal points in the new subsets is maximized at every level. This is illustrated with the 16-QAM signal set and a binary partition chain (split each set into two subsets at each level) in Figure 3.10.

Note that the partitioning can be continued until there is only one signal left in each subset. In such a way, by following the partition path,

---

[6] The notion of distance spectrum of a trellis code will be introduced and discussed in Chapter 5.

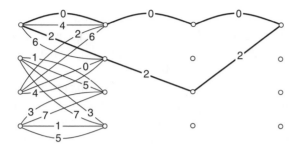

Figure 3.9 Four-state 8-PSK trellis code with parallel transitions.

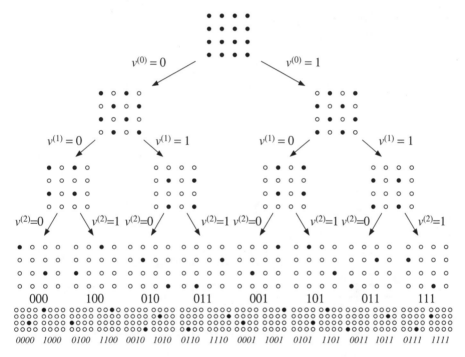

Figure 3.10 Set partitioning of a 16-QAM signal set into subsets with increasing minimum distance. The final partition level used by the encoder in Figure 3.11 is the fourth level, that is, the subsets with two signal points each.

a "natural" binary label can be assigned to each signal point. The natural labeling of the 8-PSK signal set in Figure 3.7 (M-PSK in general) can also be generated in this way. This method of partitioning a signal set is called *set partitioning* with increasing intrasubset distances. The idea is to use these constellations for codes with parallel transitions.

An alternative way of generating parallel transitions is to choose not to encode all the input bits in u_r. Using the encoder in Figure 3.11 with a 16-QAM constellation; for example, the first information bit $u_r^{(3)}$ is not encoded and the output signal of the FSM selects now a subset rather than a signal point (here one of the subsets at the fourth partition level in Figure 3.10). The uncoded bit(s) select the actual signal point within the subset. Analogously, then, the encoder now has to be designed to maximize the minimum interset distances of sequences, since it cannot influence the signal point selection within the subsets. The advantage of this strategy is that the same encoder can be used for all signal constellations with the same intraset distances at the final partition level, in particular for all signal constellations that are "expanded" versions of each other, such as 16-QAM, 32-cross, 64-QAM, etc.

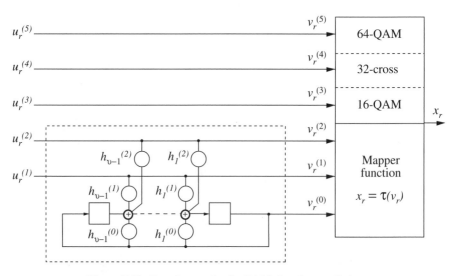

Figure 3.11 Generic encoder for QAM signal constellations.

Figure 3.11 shows such a generic encoder that maximizes the minimum interset distance between sequences, and it can be used with all QAM-based signal constellations. Only the two least significant information bits affect the encoder FSM. All other information bits cause parallel transitions. Table 3.3 shows the coding gains achievable with such an encoder structure. The gains when going from 8-PSK to 16-QAM are most marked since rectangular constellations have a somewhat better power efficiency than constant-energy constellations.

As in the case of 8-PSK codes, efforts have been made to improve the distance spectrum of a code, in particular to minimize the bit error

TABLE 3.3   Connectors and Gains of Maximum Free-Distance QAM Trellis Codes

Number of states	Connecters				Asymptotic coding gain		
	$h^{(0)}$	$h^{(1)}$	$h^{(2)}$	d^2_{free}	16-QAM/ 8-PSK (dB)	32-cross/ 16-QAM (dB)	64-QAM/ 32-cross (dB)
4	5	2	—	4.0	4.4	3.0	2.8
8	11	2	4	5.0	5.3	4.0	3.8
16	23	4	16	6.0	6.1	4.8	4.6
32	41	6	10	6.0	6.1	4.8	4.6
64	101	16	64	7.0	6.8	5.4	5.2
128	203	14	42	8.0	7.4	6.0	5.8
256	401	56	304	8.0	7.4	6.0	5.8
512	1001	346	510	8.0	7.4	6.0	5.8

SOURCE: The codes in this table were presented by Ungerböck in [4].

multiplicity of the first few spectral lines. Some of the improved codes using a 16-QAM constellation are listed in Table 3.4 together with the original codes from Table 3.3, labelled "Ung." The improved codes, taken from [6], use the signal mapping shown in Figure 3.12. Note also that the input line $u_r^{(3)}$ is also fed into the encoder for these codes.

TABLE 3.4   Original 16-QAM Trellis Code and Improved Trellis Codes Using a Nonstandard Mapping

2^v	$h^{(0)}$	$h^{(1)}$	$h^{(2)}$	$h^{(3)}$	d^2_{free}	$A_{d_{free}}$	$B_{d_{free}}$
Ung 8	11	2	4	0	5.0	3.656	18.313
8	13	4	2	6	5.0	3.656	12.344
Ung 16	23	4	16	0	6.0	9.156	53.5
16	25	12	6	14	6.0	9.156	37.594
Ung 32	41	6	10	0	6.0	2.641	16.063
32	47	22	16	34	6.0	2	6
Ung 64	101	16	64	0	7.0	8.422	55.688
64	117	26	74	52	7.0	5.078	21.688
Ung 128	203	14	42	0	8.0	36.36	277.367
128	313	176	154	22	8.0	20.328	100.031
Ung 256	401	56	304	0	8.0	7.613	51.953
256	417	266	40	226	8.0	3.273	16.391

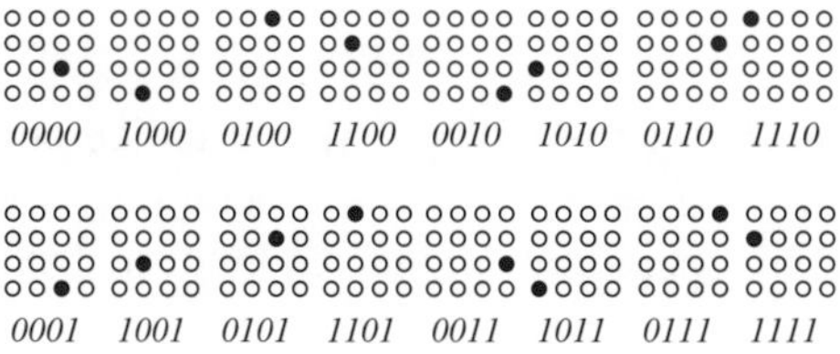

Figure 3.12 Mapping used for the improved 16-QAM codes.

3.5   LATTICES

Soon after the introduction of trellis codes and the idea of set partitioning, it was realized that certain signal constellations and their partitioning could be elegantly described in many situations by lattices. This formulation is a particularly convenient tool in the discussion of multidimensional trellis codes. (The two-dimensional complex constellations discussed so far are not considered multidimensional.) We begin by defining a lattice.

DEFINITION 3.3
An N-dimensional lattice Λ is the set of all points

$$\Lambda = \{x\} = \{i_1 b_1 + \cdots + i_N b_N\}, \tag{3.5}$$

where x is an m-dimensional row vector (point) in $\mathbf{R}^m$, $b_1, \ldots, b_N$ are N linearly independent basis vectors in $\mathbf{R}^m$, and $i_1, \ldots, i_N$ range through all integers.

A lattice is therefore something similar to a vector space, where the coefficients are restricted to be integers. Lattices have been studied in mathematics for many decades, especially in the sphere-packing problem, the covering problem, or the quantization problem. The sphere-packing problem asks the question: "What is the densest way of packing together a large number of equal-sized spheres?" The covering problem asks for the least dense way to cover space with equal overlapping spheres, and the quantization problem addresses the problem of placing points in space so that the average second moment of their Voronoi cells is as small as possible. A comprehensive treatise on lattices is given by Conway and Sloane [10]. We will use only a few results from lattice theory in our study of mapper functions.

Not surprisingly, operations on lattices can conveniently be described by matrix operations. To this end let

$$\mathbf{M} = \begin{pmatrix} b_1 \\ b_2 \\ \vdots \\ b_N \end{pmatrix} \tag{3.6}$$

be the *generator matrix* of the lattice. All the lattice points can then be generated by $i\mathbf{M}$, where $i = (i_1, \ldots, i_N)$. Operations on lattices can now easily be described by operations on the generator matrix. An example is $R\mathbf{M}$, where

$$
R = \begin{pmatrix}
1 & -1 \\
1 & 1 \\
& & 1 & -1 \\
& & 1 & 1 \\
& & & & \ddots & \ddots \\
& & & & & & 1 & -1 \\
& & & & & & 1 & 1
\end{pmatrix}.
\tag{3.7}
$$

R is an operation that rotates and expands pairs of coordinates, called the *rotation operator*, and is important for our purpose.

Two parameters of lattices are important, the minimum distance between points $d_{\min}$ and the number of nearest neighbors, lucidly described as the *kissing number* τ of the lattice. This is because if we fill the m-dimensional Euclidean space with m-dimensional spheres of radius $d_{\min}/2$ centered at the lattice points, there are exactly τ spheres that touch (kiss) any given sphere. The density Δ of a lattice is the proportion of the space occupied by these touching spheres. Also, the fundamental volume $V(\Lambda)$ is the N-dimensional volume per lattice point; that is, if we partition the total space into regions of equal size $V(\Lambda)$, each such region is associated with one lattice point.

Lattices have long been used in working with the sphere-packing problem, which attempts to pack spheres in m dimensions, such that as much of the space as possible is covered by spheres. Lattices are being investigated for the problem, because if we find a "locally" dense packing, it is also "globally" dense due to the linearity of the lattice.

A popular lattice is Z^N, the *cubic lattice*, consisting of all N-tuples with integer coordinates. Its minimum distance is $d_{\min} = 1$ and its kissing number is $\tau = 2N$. Trivially, Z^1 is the densest lattice in one dimension.

Another interesting lattice[7] is D_N, the *checkerboard lattice*, whose points consist of all points whose integer coordinates sum to an even number. It can be generated from Z^N by casting out that half of all the points that have an odd integer sum. In doing so we have increased $d_{\min}$ to $\sqrt{2}$ and the kissing number to $\tau = 2N(N-1)$, for $N \geq 2$, and have obtained a much denser lattice packing, which can be seen as follows. Since we need to normalize $d_{\min}$ in order to compare packings, let us shrink D_N by $(1/\sqrt{2})^N$. This puts $\sqrt{2}^N/2$ as many points as Z^N into the same volume. (The denominator equals 2 since we eliminated half the points). Therefore D_N is $2^{N/2-1}$ times as dense as Z^N. Note that D_2 has the same density as

---

[7] The reason for the different notation of Z^N and D_N is that $Z^N = Z \times Z \times \cdots \times Z$, the Cartesian product of one-dimensional lattices Z, which D_N is not.

Z^2, and it is in fact the rotated version of the latter; that is, $D_2 = RZ^2$. D_4, the *Schläfli lattice*, is the densest lattice packing in four dimensions. We will return to D_4 in our discussion of multidimensional trellis codes.

To describe signal constellations by lattices, we have to shift and scale the lattice. To obtain the rectangular constellations Q from Z^2, for example, we set $Q = c(Z^2 + \{\frac{1}{2}, \frac{1}{2}\})$, where c is an arbitrary scaling factor, usually chosen to normalize the average symbol energy to unity, and the shift by $(\frac{1}{2}, \frac{1}{2})$ centers the lattice. As can be seen from Figure 3.13, such a shifted lattice can be used as a template for all QAM constellations.

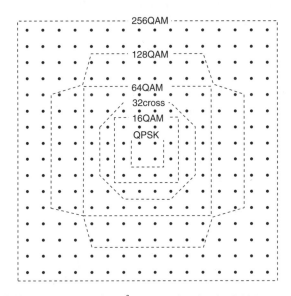

Figure 3.13 The integer lattice Z^2 as a template for the QAM constellations.

To extract a finite signal constellation from the lattice, we introduce a boundary and choose only points inside the boundary. This boundary shapes the signal constellation and affects the average power used in transmitting signal points from such a constellation. The effect of this shaping is summarized in the shaping gain γ_s.

If we consider the rectangular constellations (see Figure 3.13), the area of the square and the cross is given by $V(Z^2)2^n$, where 2^n is the number of signal points. The average energy $\overline{E_s}$ of these two constellations can be calculated by an integral approximation for large numbers of signal points, and we obtain $\overline{E_s} = \frac{2}{3}2^n$ for the cross and $\overline{E_s} = \frac{31}{48}2^n$ for the cross shape. That is, the cross constellation is $\frac{31}{32}$ or $\gamma_s = 0.14$ dB more energy efficient than the rectangular boundary.

The best enclosing boundary would be a circle, especially in higher dimensions. It is easy to see that the energy saving of an N-dimensional spherical constellation over an N-dimensional rectangular constellation is proportional to the inverse ratio of their volumes. This gain can be calculated as [11]

$$\gamma_s = \frac{\pi(N+2)}{12}\left(\left(\frac{N}{2}\right)!\right)^{-2/N}, \qquad N \text{ even.} \tag{3.8}$$

We see that in the limit $\lim_{N\to\infty}\gamma_s = \pi e/6$, or 1.53 dB. To obtain that gain, however, the constituent two-dimensional signal points in our N-dimensional constellation will not be distributed uniformly. In fact, they will tend toward a Gaussian distribution.

The fundamental coding gain[8]

$$\gamma(\Lambda) = d_{\min}^2/V(\Lambda)^{2/N} \tag{3.9}$$

and relates the minimum squared Euclidean distance to the fundamental volume per two dimensions. This definition is meaningful since the volume per two dimensions is directly related to the signal energy, and thus $\gamma(\Lambda)$ expresses an asymptotic coding gain. For example, from the preceding discussion

$$\gamma(D_N) = \frac{2}{(2V(\Lambda))^{2/N}} = 2^{1-2/N}\gamma(Z^N), \tag{3.10}$$

$\gamma(Z^N) = 1$ shall be used as reference henceforth. Definition (3.9) is also invariant to scaling, as one would expect. Thus, for example, $\gamma(D_4) = 2^{1/2}$; that is, D_4 has a coding gain of $2^{1/2}$ or 1.51 dB over Z^4.

Partitioning of signal sets can now be discussed much more conveniently with the help of the lattice and is applicable to all signal constellations derived from that lattice. The binary set partitioning in Figure 3.10 can be derived from the binary lattice partition chain

$$Z^2/RZ^2/2Z^2/2RZ^2 = Z^2/D_2/2Z^2/2D_2. \tag{3.11}$$

Every level of the partition in (3.11) creates a sublattice Λ' of the original lattice Λ as shown in Figure 3.14. The remaining points, if we delete Λ' from Λ, are a shifted version Λ'_s of Λ', called its coset. (There are p cosets in a p-ary partition.) Λ' and Λ'_s make up Λ; that is, $\Lambda = \Lambda' \bigcup \Lambda'_s$. (For example, $Z^2 = RZ^2 \bigcup (RZ^2 + (1,0))$.) The partition chain (3.11)

---

[8] The fundamental coding gain is related to the center density δ used in [10] by $\gamma(\Lambda) = 4\delta^{2/N}$. The center density of a lattice is defined as the number of points per unit volume if the touching spheres have unit radius.

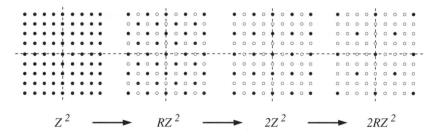

Figure 3.14 Illustration of the binary partition chain (3.11).

generates $2^3 = 8$ cosets of $2RZ^2$, and each coset is a translate of the final sublattice.

A coset of a lattice Λ is denoted by $\Lambda + \mathbf{c}$, where $\mathbf{c}$ is some constant vector that specifies the coset. Note that if $\mathbf{c}$ is in Λ, $\Lambda + \mathbf{c} = \Lambda$. Mathematically speaking, we have defined an equivalence relation of points; that is, all points equivalent to $\mathbf{c}$ modulo Λ are in the same coset $\Lambda + \mathbf{c}$.

If we now have a lattice partition Λ/Λ' and we choose $\mathbf{c}$ such that $\Lambda' + \mathbf{c} \in \Lambda$ (note: not Λ'), then every element in Λ can be expressed as

$$\Lambda = \Lambda' + [\Lambda/\Lambda'], \tag{3.12}$$

where $[\Lambda/\Lambda'$ is just a fancy way of writing the set of all such vectors $\mathbf{c}$. Equation (3.12) is called the coset decomposition of Λ in terms of cosets (translates) of the lattice Λ' and the number of such different cosets is the order of the partition, denoted by $|\Lambda/\Lambda'|$. As an example consider the decomposition of Z^2 into $Z^2 = RZ^2 + \{(0, 0), (0, 1)\}$. Analogously an entire partition chain can be defined; for example, for $\Lambda/\Lambda'/\Lambda''$ we may write $\Lambda = \Lambda'' + [\Lambda'/\Lambda''] + [\Lambda/\Lambda']$—that is, every element in Λ can be expressed as an element of Λ'' plus a shift vector from $[\Lambda'/\Lambda'']$ plus a shift vector from $[\Lambda/\Lambda']$. Again, for example, $Z^2 = 2Z^2 + \{(0, 0), (1, 1)\} + \{(0, 0), (0, 1)\}$ (see Figure 3.14).

The lattice partitions discussed here are all binary partitions; that is, at each level the lattice is split into two cosets as in (3.11). The advantage of binary lattice partitions is that they naturally map into binary representations. Binary lattice partitions will be discussed more in Chapter 7.

Let us then consider a chain of K binary lattice partitions; that is,

$$\Lambda_1/\Lambda_2/\cdots\Lambda_K, \tag{3.13}$$

for which (3.11) may serve as an example of a chain of four lattices. There are then 2^K cosets of Λ_K whose union makes up the original lattice Λ_1.

Each such coset can now conveniently be identified by a K-ary binary vector $v = (v^{(1)}, \ldots, v^{(K)})$; that is, each coset is given by Λ_K shifted by

$$\mathbf{c}(v) = \sum_{i=1}^{K} v^{(i)} \mathbf{c}^{(i)}, \qquad (3.14)$$

where $\mathbf{c}^{(i)}$ is an element of Λ_i but not of Λ_{i+1}. The two vectors $\{0, \mathbf{c}^{(i)}\}$ are then two coset representatives for the cosets of Λ_{i+1} in the binary partition $\Lambda_i / \Lambda_{i+1}$. (Compare the example for the partition $Z^2/2Z^2$, where $\mathbf{c}^{(1)} = (0, 1)$ and $\mathbf{c}^{(2)} = (1, 1)$; see also Figure 3.15). Generalizing to the chain of length K, the cosets of Λ_K in the partition chain Λ/Λ_K are given by the linear sum (3.14).

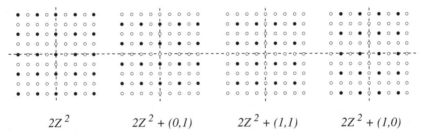

$$2Z^2 \qquad\qquad 2Z^2 + (0,1) \qquad\qquad 2Z^2 + (1,1) \qquad\qquad 2Z^2 + (1,0)$$

Figure 3.15 The four cosets of $2Z^2$ in the partition $Z^2/2Z^2$.

In fact, the partition chain (3.11) is the basis for the generic encoder in Figure 3.11 for all QAM constellations. With the lattice formulation, we wish to describe our trellis encoder in the general form of Figure 3.16. The FSM encodes k bits, the mapper function selects one of the cosets of the final sublattice of the partition chain, and the $n-k$ uncoded information bits select a signal point from that coset. The encoder can only affect the choice of cosets and therefore needs to be designed to maximize the minimum distance between sequences of cosets, where the minimum distance between two cosets Λ and Λ' is defined as $\min_{x \in \Lambda', x' \in \Lambda'_s} |x, x'|^2 = \min_{x' \in \Lambda'_s} |x'|^2$ (taking the origin as reference point; i.e., $x = 0$).

As shown in Figure 3.16 the trellis code breaks up naturally into two components. The first is a called a *coset code* and is made up of the finite-state machine and the mapping into cosets of a suitable lattice partition Λ/Λ'. The second part is the choice of the actual signal point within the chosen coset. This signal choice determines the constellation boundary and therefore shapes the signal set. It is referred to as *shaping*.

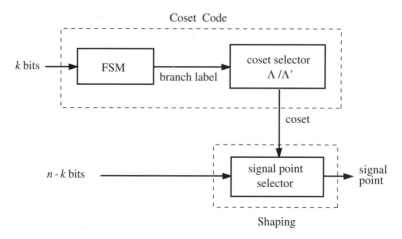

Figure 3.16 Generic trellis encoder block diagram using lattice notation.

3.6   LATTICE FORMULATION OF TRELLIS CODES

A coset code as shown in Figure 3.16 will be denoted by $C(\Lambda/\Lambda'; C)$, where C is the FSM generating the branch labels. In most known cases C is a convolutional code. The sequence of branch labels, the codewords of C, will be written as $(v_r, v_{r+1}, \ldots)$, that is, a sequence of binary $(k + m)$-tuples, where m is the number of parity bits in the convolutional code (e.g., $m = 1$ for the codes generated by the encoder in Figure 3.6). Each v_j serves as a label that selects a coset of Λ' at time r, and the coded part of the encoder selects therefore a sequence of lattice cosets. The $n - k$ uncoded bits select one of 2^{n-k} signal points from the coset Λ'_r at time r. Writing the branch labels v_j in the power series form with the delay operator D,

$$v(D) = v_r D^r + v_{r+1} D^{r+1} + \cdots, \tag{3.15}$$

we may say that a coset code maps a label sequence $v(D)$ into a sequence of cosets $c(D)$.

Since, in general, the branch labels of our trellis code are now cosets containing more than one signal point, the minimum distance of the code is $\min(d^2_{\text{free}}, d^2_{\text{min}})$, where d^2_{free} is the minimum free-squared Euclidean distance between any two output sequences and d^2_{min} is the minimum squared Euclidean distance between members of Λ', the final sublattice of Λ.

Figure 3.17 shows the generic QAM encoder in lattice form, where $v^{(0)}$ selects one of the two cosets of RZ^2 in the partition Z^2/RZ^2, $v^{(1)}$ selects

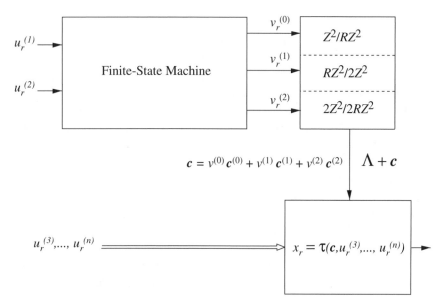

Figure 3.17 Lattice encoder for the two-dimensional rectangular constellations grown from the lattice Z^2.

one of the two cosets of $2Z^2$ in the partition $RZ^2/2Z^2$, and $v^{(2)}$ selects one of the two cosets in the partition $2Z^2/2RZ^2$ (compare also Figures 3.10 and 3.11). The final selected coset in the $Z^2/2RZ^2$ partition is given by

$$\mathbf{c} = v^{(0)}\mathbf{c}^{(0)} + v^{(1)}\mathbf{c}^{(1)} + v^{(2)}\mathbf{c}^{(2)}, \tag{3.16}$$

where $\mathbf{c}^{(0)}$ is the coset representative in the partition Z^2/RZ^2 and $\mathbf{c}^{(1)}$ and $\mathbf{c}^{(2)}$ are the coset representatives in the partitions $RZ^2/2Z^2$ and $2Z^2/2RZ^2$, respectively.

An important fact is that, contrary to the trellis code in Figure 3.11, the coset codes discussed in this section are group codes; that is, two output coset sequences $c_1(D)$ and $c_2(D)$ may be added to produce another valid output sequence $c_3(D) = c_1(D) + c_2(D)$. The reason why this is the case lies precisely in the lattice formulation. With the lattice viewpoint we have also introduced an addition operator that allows us to add lattice points or entire lattices. We also note that the lattices themselves are infinite, and there are no boundary problems. We have already remarked that a group code is trivially regular and we do not need to worry about all correct sequences and may choose any convenient one as reference.

With this new lattice description, trellis codes can now easily be classified. Ungerböck's original one- and two-dimensional pulse-amplitude-modulated (PAM) codes are based on the four-way partition $Z/4Z$ and the eight-way partition $Z^2/2RZ^2$, respectively. The one-dimensional codes are used with a rate $R = 1/2$ convolutional code and the two-dimensional codes with a rate $R = 2/3$ convolutional code with the exception of the four-state code ($\nu = 2$), which uses a rate $R = 1/2$ code.

Table 3.5 shows these codes where the two-dimensional codes are essentially a reproduction of Table 3.3. Also shown in the table is N_D, the number of nearest neighbors in the coset code. Note that the number of nearest neighbors is based on the infinite lattice and selecting a particular constellation from the lattice will reduce that number. As we discuss in more detail later, the number of nearest neighbors, and in fact the number of neighbors at a given distance in general, also affects performance, and sometimes codes with a smaller d_{free}^2 can outperform codes with a larger d_{free}^2 but more nearest neighbors.

So far we have only concerned ourselves with two-dimensional lattices as a basis for our signal constellations, but this is an arbitrary re-

TABLE 3.5 Lattice Partition, Minimum Squared Euclidean Distance, Asymptotic Coding Gain, and Number of Nearest Neighbors for the Original Ungerböck One- and Two-Dimensional Trellis Codes

Number of states	Λ	Λ'	d_{min}^2	Asymptotic coding gain (dB)	N_D
One-dimensional codes					
4	Z	$4Z$	9	3.52	8
8	Z	$4Z$	10	3.98	8
16	Z	$4Z$	11	4.39	16
32	Z	$4Z$	13	5.12	24
64	Z	$4Z$	14	5.44	72
128	Z	$4Z$	16	6.02	132
256	Z	$4Z$	16	6.02	4
512	Z	$4Z$	16	6.02	4
Two-dimensional codes					
4	Z^2	$2Z^2$	4	3.01	4
8	Z^2	$2RZ^2$	5	3.98	16
16	Z^2	$2RZ^2$	6	4.77	56
32	Z^2	$2RZ^2$	6	5.77	16
64	Z^2	$2RZ^2$	7	5.44	56
128	Z^2	$2RZ^2$	8	6.02	344
256	Z^2	$2RZ^2$	8	6.02	44
512	Z^2	$2RZ^2$	8	6.02	4

striction and multidimensional trellis codes—trellis codes using lattices of dimensions larger than 2—have been constructed and have several advantages. Multidimensional in this context is a theoretical concept, since, in practice, multidimensional signals are transmitted as sequences of one- or two-dimensional signals.

We have seen that to introduce coding we need to expand the signal set from the original uncoded signal set. The most usual and convenient way is to double the constellation size—introduce 1 bit of redundancy. Doubling reduces the minimum distance within the constellation, and this reduction has to be compensated for by the code before any coding gain can be achieved. If we use, say, a four-dimensional signal set, doubling causes the constituent two-dimensional constellations to be expanded by a factor of only $\sqrt{2}$ (half a bit of redundancy per two-dimensional constellation). The advantage is that there is less initial loss in minimum distance within the signal set. If we consider rectangular signal sets derived from Z^N, we obtain the following numbers. For two dimensions the signal set expansion costs 3 dB in minimum distance loss, for four-dimensional signals the loss is 1.5 dB, and for eight-dimensional signal sets it is down to 0.75 dB. We see that the code itself has to overcome less and less signal set expansion loss. Another point in favor of multidimensional codes is that linear codes with $90°$ phase invariance can be constructed. This will be explored in more detail in Section 3.7.

Let us consider codes over four-dimensional rectangular signal sets as an example. The binary partition chain we use is

$$Z^4/D_4/RZ^4/RD_4/2Z^4/2D_4/ \cdots \quad (3.17)$$

with minimum distances

$$1/\sqrt{2}/\sqrt{2}/2/2/2\sqrt{2}/ \cdots. \quad (3.18)$$

The FSM generating the code trellis, usually a convolutional code, is now designed to maximize the intraset distance of sequences. No design procedure for good codes is known to date, and computer searches are usually carried out, either exhaustively or with heuristic selection and rejection rules. Note, however, that the computer search needs to optimize only the coset code, which is linear, and therefore we can choose $v(D) = 0$ as the reference sequence. Now we have a simple mapping of v_r into a distance increment d_i^2, and $d_{\text{free}}^2 = \min_{\mathbf{c}(D),\mathbf{c}'(D)} \left\| \mathbf{c}(D) - \mathbf{c}'(D) \right\|^2 = \min_{v(D)} \sum_r d_r^2(v_r)$, where $d_r^2(v_r) = \min_{x \in \Lambda_s(v_r)} \|x\|^2$ is the distance from the origin of the closest point in the coset $\Lambda_s(v_r)$ specified by v_r.

Figure 3.18 illustrates the partitioning tree analogously to Figure 3.10 for the four-dimensional eight-way partition $\Lambda/\Lambda' = Z^4/RD_4$. The cosets at each partition are given as unions of two two-dimensional cosets; that is, $D_2 \times D_2 \cup \overline{D_2} \times \overline{D_2}$ is the union of the Cartesian product of D_2 with D_2 and $\overline{D_2}$ with $\overline{D_2}$, where $\overline{D_2} = D_2 + (0, 1)$ is the second coset.

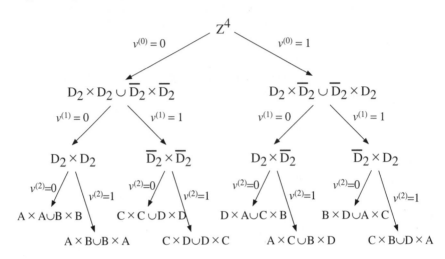

Figure 3.18 Eight-way lattice partition tree of Z_4. The constituent two-dimensional cosets are $A = 2Z^2$, $B = 2Z^2 + (1, 1)$, $C = 2Z^2 + (0, 1)$, and $D = 2Z^2 + (1, 0)$ as depicted in Figure 3.15.

Figure 3.19 shows a 16-state encoder using a rate $R = 2/3$ convolutional code to generate a trellis with $d^2_{\text{free}} = d^2_{\text{min}} = 4$; that is, the parallel transitions are those producing the minimum distance. Since the final lattice $\Lambda' = RD_4$ is a rotated version of the checkerboard lattice D_4, we immediately also know the number of nearest neighbors at $d^2_{\text{min}} = 4$, which

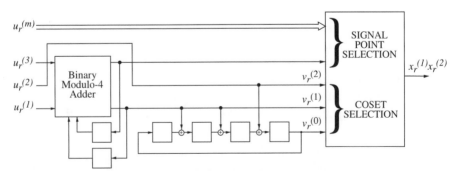

Figure 3.19 Sixteen-state four-dimensional trellis code using the eight-way partition shown in Figure 3.18. The differential encoder on the input bits is used to make the code rotationally invariant.

is $N_D = 24$. The addition of the differential encoder allows the code to be made rotationally invariant to $90°$ phase rotations. This particular encoder will be discussed further in Section 3.7 on rotational invariance.

One philosophy, brought forth by Wei [12], is to choose a lattice partition Λ/Λ' where Λ' is a denser lattice than Λ. A dense lattice Λ' increases the minimum distance, and therefore the asymptotic coding gain, but this also increases the number of nearest neighbors. Another advantage is that this philosophy simplifies code construction since codes with fewer states may be used to achieve the same gain.

Table 3.6 summarizes the best-known multidimensional trellis codes found to date. The index in the last column refers to the source of the

TABLE 3.6   Best Multidimensional Trellis Codes Based on Binary Lattice Partitions of 4, 8, and 16 Dimensions

Number of states	Λ	Λ'	d_{min}^2	Asymptotic coding gain (dB)	N_D	*Source*
Four-dimensional codes						
8	Z^4	RD_4	4	4.52	44	W
16	D_4	$2D_4$	6	4.77	152	C-S
64	D_4	$2D_4$	8	5.27	828	C-S
16	Z^4	RD_4	4	4.52	12	W
32	Z^4	$2Z^4$	4	4.52	4	W
64	Z^4	$2D_4$	5	5.48	72	W
128	Z^4	$2D_4$	6	6.28	728	U
Eight-dimensional codes						
16	Z^8	E_8	4	5.27	316	W
32	Z^8	E_8	4	5.27	124	W
64	Z^8	E_8	4	5.27	60	W
128	Z^8	RD_8	4	5.27	28	U
32	RD_8	RE_8	8	6.02		W
64	RD_8	RE_8	8	6.02	316	W
128	RD_8	RE_8	8	6.02	124	W
8	E_8	RE_8	8	5.27	764	C-S
16	E_8	RE_8	8	5.27	316	C-S
32	E_8	RE_8	8	5.27	124	C-S
64	E_8	RE_8	8	5.27	60	C-S
Sixteen-dimensional codes						
32	Z^{16}	H_{16}	4	5.64		W
64	Z^{16}	H_{16}	4	5.64	796	W
128	Z^{16}	H_{16}	4	5.64	412	W

SOURCE: The source column indicates the origin of the code; that is, codes marked *U* were found in [4], those marked *W* in [12], and those marked *C-S* in [15]. The Gosset lattice E_8 and the lattice H_{16} are discussed in Chapter 7.

code. Note that the table also includes eight-dimensional codes, mainly pioneered by Wei [12] and Calderbank and Sloane [13, 15]. We have discussed the construction and classification of multidimensional codes, but we have said little about the motivation to use more than two dimensions. Two-dimensional signal constellations seem to be a natural choice since bandpass double-sideband-modulated signals naturally generate a pair of closely related dimensions, the in-phase and quadrature channels. The motivation to go to higher-dimensional signal sets is not quite so obvious, but there are two very important reasons. The first is the smaller constellation expansion factor discussed previously. The second is that multidimensional codes can be made invariant to phase rotations by using linear FSMs.

All the lattices discussed (D_2, D_4, E_8, H_{16}), are sublattices of the Cartesian product lattice Z^N. Signal constellations derived from these lattices are particularly useful since a modem built for Z^2, that is, QAM constellations, can easily be reprogrammed to accommodate the lattices without much effort. We simply disallow the signal points in Z^N that are not points in the sublattice in question. Table 3.7 shows these useful sublattices, together with some of their parameters:

TABLE 3.7   Popular Sublattices of Z^N, Their Minimum Distances $d^2_{\min}$, and Their Kissing Numbers (Nearest Neighbors)

Lattice	Kissing numbers (nearest neighbors)	$d^2_{\min}$	Fundamental coding gain (γ_Λ)
D_N (Checkerboard)	$2N(N-1)$	2	$2^{1-2/N}$
E_8 (Gosset)	240	4	2
H_{16}		4	
Λ_{16} (Barnes-Wall)	4,320	8	$2^{3/2}$
Λ_{24} (Leech)	196,560	8	4
D_{32} (Barnes-Wall)	208,320	16	4

3.7 ROTATIONAL INVARIANCE

With ideal coherent detection of a DSB-SC signal (Figure 2.7), the absolute carrier phase of the transmitted signal needs to be known at the receiver to ensure the orientation of the signal constellation is correct. Let us then assume that the received signal has a phase offset by an angle ϕ; that is, the received signal, ignoring noise, is given by (equation (2.31))

$$s_0(t) = x(t)\sqrt{2}\cos(2\pi f_0 t + \phi). \tag{3.19}$$

Let us assume for illustration that we are using an M-PSK signal set. We can then write (3.19) as

$$s_0(t) = \sqrt{2}\cos(2\pi f_0 t + \phi_m(t) + \phi), \qquad (3.20)$$

where $\phi_m(t)$ is the time-varying data phase. For example, for QPSK $\phi_m(t)$ can assume the angles $\pi/2$, π, $3\pi/2$, and 2π. The carrier phase-tracking loop, usually a *phase-locked loop* (PLL) circuit, first needs to eliminate the data-dependent part of the phase. In the case of the QPSK signal, this can be accomplished by raising $s_0(t)$ to the fourth power. This produces a spectral line at $4f_0$, given by $4\cos(2\pi 4 f_0 + 4\phi_m(t) + 4\phi)$. But $4\phi_m(t)$ is always a multiple of 2π for QPSK modulation, and the fourth-power spectral line is now free of the data-dependent phase changes, given by $4\cos(2\pi 4 f_0 + 4\phi)$. A PLL is now used to track the phase, and the local oscillator signal is generated by dividing the frequency of the tracking signal by 4.

We see that phase offsets of multiples of $\pi/2$ cannot be identified by this procedure, leaving the carrier phase recovery system with a phase ambiguity of multiples of $\pi/2$. These phase ambiguities have the undesirable effect that the received signal constellations may be rotated versions of the transmitted constellations, and the trellis decoder usually cannot decode properly if the constellations are rotated. This necessitates that all possible phase rotations of the constellation have to be tried until the correct one is found. It is therefore desirable to design trellis-coded modulation (TCM) systems such that they have as many phase invariances as possible—that is, rotation angles that do not affect the decoder operation. This will also ensure more rapid resynchronization after temporary signal loss or phase jumps.

In decision-directed carrier phase acquisition and tracking circuits, this rotational invariance is even more important, since without proper output data the tracking loop is in an undriven random-walk situation, from which it may take a long time to recover. This situation also can be avoided by making the trellis code invariant to constellation rotations.

A trellis code is called *rotationally invariant* with respect to the rotation of constituent constellation by an angle ϕ if the decoder can correctly decode the transmitted information sequence when the local oscillator phase differs from the carrier phase by ϕ. Naturally the phase ambiguity angle of the recovery circuit is restricted to be a rotational symmetry angle of the employed signal constellation, that is, an angle that rotates the signal constellation into itself. If $\mathbf{x}$ is a sequence of coded symbols, denote by $\mathbf{x}^\phi$ the symbol sequence obtained by rotating each symbol x_r by the angle ϕ. We now have the following definition.

DEFINITION 3.4

A TCM code is rotationally invariant with respect to a rotation by an angle ϕ if $\mathbf{x}^\phi$ is also a code sequence for all valid code sequences $\mathbf{x}$.

This definition is rather awkward to test or work with. But we can translate it into conditions on the transitions of the code trellis. Assume that there is a total of S states in S and that there are P subsets that result from set-partitioning the original constellation. Assume further that the partitioning is done such that, for each possible phase rotation, each subset rotates into itself or another subset; that is, the set of subsets is invariant under these rotations. This latter point is automatically true for one- and two-dimensional constellations [12] (see, e.g., Figure 3.10) if we use the type of lattice partitioning discussed in the previous sections. With these assumptions we now may state the following [12, 14].

THEOREM 3.2

A TCM code is rotationally invariant with respect to a rotation by an angle ϕ if there exists a (bijective) function $f_\phi : S \mapsto S$ with the following properties: For each transition from a state i to a state j, denote the associated subset by A. Denote by B the subset obtained when A is rotated by the angle ϕ. Then B is the subset associated with the transition from state $f_\phi(i)$ to state $f_\phi(j)$.

PROOF

The situation in Theorem 3.2 is illustrated in Figure 3.20. From there it is easy to see that, if we concatenate successive trellis sections, for each path $i, j, k, \ldots$, there exists a valid path $f_\phi(i), f_\phi(j), f_\phi(k), \ldots$, of rotated symbols through the trellis.

There are now two components to making a trellis code transparent to a phase rotation of the signal constellation. First, the code must be rotationally invariant; that is, the decoder will still find a valid code sequence after rotation. Second, the rotated sequence needs to map back into the same information bits as the unrotated sequence. This is achieved by differentially encoding the information bits, as illustrated below.

It was impossible to achieve rotational invariance of a code with a group-trellis code in conjunction with two-dimensional signal constellations [16–18], so nonlinear trellis codes were investigated [19, 20]. Figure 3.21 shows such a nonlinear eight-state trellis code for use with QAM constellations, illustrated for use with the 32-cross constellation to transmit 4 bits/symbol, as shown in Figure 3.22. This code was adopted in the CCITT V.32 Recommendation [21] and provides an asymptotic coding gain of 4 dB. The three leading bits (lightface type in the figure) are not

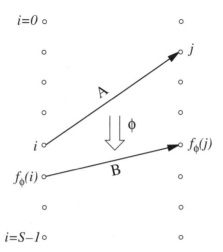

Figure 3.20 Illustration of Theorem 3.2.

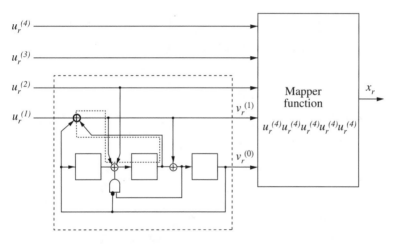

Figure 3.21 Nonlinear rotationally invariant trellis code with eight states for
           QAM constellations.

affected by any 90° rotation of the constellation. Only the last two bits
(boldface type) are affected and need to be encoded differentially, as is
done in the encoder. The constellation is partitioned into the eight subsets
$D_0, \ldots, D_7$. Note that successive 90° phase rotations rotate $D_0 \rightarrow D_1 \rightarrow$
$D_2 \rightarrow D_3 \rightarrow D_0$ and $D_4 \rightarrow D_5 \rightarrow D_6 \rightarrow D_7 \rightarrow D_4$.

Equipped with this information we may now study a section of the trellis
diagram of this nonlinear, rotationally invariant trellis code. This trellis
section is shown in Figure 3.23, together with the state correspondence

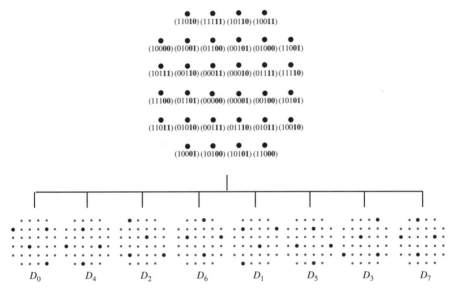

Figure 3.22 Example 32-cross constellation for the nonlinear trellis code of
Figure 3.21. The set partitioning of the 32-cross constellation into
the eight subsets $D_0, \ldots, D_7$ is also illustrated [4].

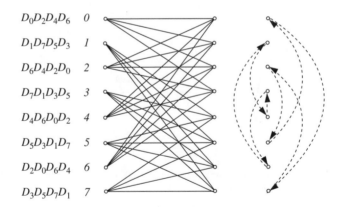

Figure 3.23 Trellis section of the nonlinear eight-state trellis code. The subset
labeling is such that the first signal set is the one on the top branch
and the last signal set is on the lowest branch throughout all the
states.

function $f_\phi(i)$, for successive 90° rotations of the signal constellations
(see Theorem 3.2). It is relatively easy to see that if we take an example
sequence D_4, D_0, D_3, D_6, D_4 and rotate it 90° into D_5, D_1, D_0, D_7, D_5,
we obtain another valid sequence of subsets. Careful checking reveals that

the function required in Theorem 3.2 exists, and the corresponding state relations are illustrated in Figure 3.23.

This then takes care of the rotational invariance of the code with respect to phase rotations of multiples of 90°. However, the bit $v_r^{(1)}$ changes its value through such rotations. This is where the differential encoding comes into play. In Figure 3.21 this function is realized by the XOR gate on the input line $u_r^{(1)}$ and the second delay cell; that is, the differential encoder is integrated into the trellis encoder. This integrated differential encoder is indicated by the dotted loop in Figure 3.21. The coding gain of this code is 4 dB and the number of nearest neighbors is 16 [4].

Standards V.32 and V.33 [22] use the trellis code in Figure 3.24. The code is an eight-state trellis code with 90° phase invariance. The V.32 standard operates at duplex rates of up to 9600 bit/s, at a symbol rate of 2400 baud using the rotated cross-constellation of Figure 3.25. We leave it as an exercise to show that the V.32 code is rotationally invariant to phase offsets of multiples of $\pi/2$.

The V.33 recommendation allows for data rates up to 14,400 bit/s. It is designed to operate over point-to-point, four-wire leased telephone-type circuits. It uses the same encoder as V.32 but with an expanded 128-point cross-signal constellation. It, too, is invariant to phase rotations of multiples of $\pi/2$.

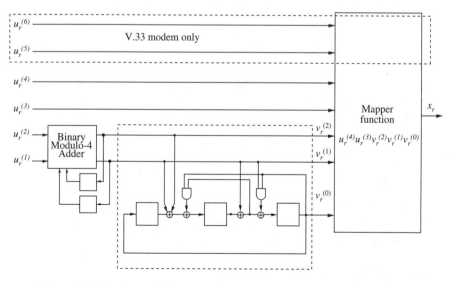

Figure 3.24 Nonlinear eight-state trellis code used in the V.32 and V.33 recommendations.

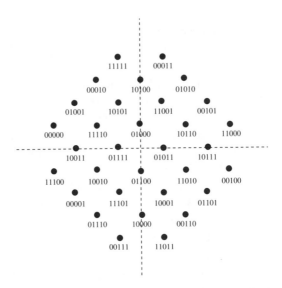

Figure 3.25 A 32-point rotated cross-signal constellation used in the V.32 recommendation for a 9600 bit/s voiceband modem.

More commonly, the differential encoder is used externally, as in Figure 3.19. The signal set partition chain for that encoder was given in Figure 3.18. A 90° phase rotation will cause the following signal set changes:[9] $A \times A \cup B \times B \to C \times C \cup D \times D \to A \times A \cup B \times B, A \times B \cup B \times A \to C \times D \cup D \times C \to A \times B \cup B \times A, D \times A \cup C \times B \to B \times D \cup A \times C \to D \times A \cup C \times B$, and $A \times C \cup B \times D \to C \times B \cup D \times A$ (see Figures 3.15 and 3.18). A portion of the trellis of this code is shown in Figure 3.26.

The state correspondences are worked out by checking through Theorem 3.2, from where we find that $f_{90°}(0) = 4$, $f_{90°}(4) = 0$, etc., as indicated in Figure 3.26 by the arrows.

Note that, due to the multidimensional nature of the signal set, 180° phase rotations map the subsets back onto themselves. This is the reason why a linear code suffices, since only one phase ambiguity has to be eliminated. The remaining ambiguities can be taken care of by differential encoding within the subsets.

Wei [14] and Pietrobon et al. [17] tackle the problem of designing rotationally invariant trellis codes for M-PSK signal constellations. This is somewhat more complicated, since the number of phase ambiguities of the constellations are larger than for QAM constellations [23]. The basic philosophy, however, remains the same.

---

[9] Recall that the lattice is shifted by $(\frac{1}{2}, \frac{1}{2})$ in order to obtain the constellation.

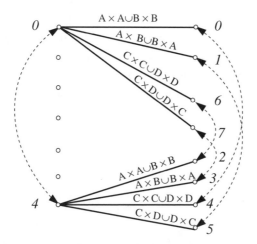

Figure 3.26 Portion of the trellis section of the linear 16-state trellis code, which achieves 90° rotational invariance.

The fastest modems can achieve a data transmission rate of up to 28,800 bit/s over the public-switched telephone network (PSTN). This is roughly equal to the information-theoretic capacity of an additive white Gaussian noise (AWGN) channel with the same bandwidth. The signal processing techniques that the modems can use adaptively and automatically are selection of the transmission band, selection of the trellis code, constellation shaping, precoding and preemphasis for equalization, and signal set warping. These techniques are summarized in [24].

These modems use trellis coding and are described in the standard recommendation V.34, also nicknamed V.fast [25]. The V.fast modems use several symbol rates (from 2400 baud up to 3429 baud) and a four-dimensional signal constellation. The constituent two-dimensional signal constellations are derived from a rectangular signal constellation and have 240 signal points each. V.34 allows the use of three different trellis codes, a 16-state, a 32-state, and a 64-state code.

3.8   GEOMETRIC UNIFORMITY[10]

The lattice viewpoint of the last sections allowed us to simplify our treatment of trellis codes considerably by making possible the decomposition of the code into a coset code and a signal selector. In this section we wish to

---

[10] This section makes use of some basic group-theoretic concepts. It can be skipped at a first reading without disrupting continuity.

generalize these notions. To do this we use the language of geometry. The unique property of the lattice that allowed the elegant treatment of trellis codes was its linearity—that is, the property that the lattice looked identical from every lattice point. Thus, appropriate operations such as rotations or translations of the lattice, which moved one signal point into another, left the lattice unchanged. Such operations belong to the class of *isometries*, defined in

DEFINITION 3.5

An isometry is an operation $T(x)$, where $x, T(x) \in \mathbf{R}^m$, such that

$$\left\| T(x) - T(x') \right\|^2 = \left\| x - x' \right\|^2. \tag{3.21}$$

That is, an isometry leaves the Euclidean distance between two points unchanged. Typical examples of isometries are rotations and translations. In fact every isometry in $\mathbf{R}^m$ can be decomposed into a sequence of translations, rotations, and reflections.

Isometries are a tool for our treatment of signal sets. Two signal sets S_1 and S_2 are said to be *congruent* if there exists an isometry $T(x)$ that, when applied to S_1, produces S_2; that is, $T(S_1) = S_2$. Furthermore, an isometry that, when applied to a signal set S, reproduces S is called a *symmetry* of S. Symmetries are therefore isometries that leave a signal set invariant. For example, for a QPSK signal set, the symmetries are the rotations of $0°, 90°, 180°$, and $270°$, the two reflections by the main axes, and the two reflections about the $45°$ and $135°$ lines. These symmetries form a group under composition, called the *symmetry group* of S.

A geometrically uniform signal set is now defined by

DEFINITION 3.6

A signal set S is geometrically uniform if for any two points $x_i, x_j \in S$ there exists an isometry $T_{i \to j}$ that carries x_i into x_j and leaves the signal set invariant; i.e.,

$$T_{i \to j}(x_i) = x_j, \qquad T_{i \to j}(S) = S. \tag{3.22}$$

Hence a signal set is geometrically uniform if we can take an arbitrary signal point and, through application of isometries $T_{i \to j}$, generate the entire signal constellation S. Another way of saying this is that the symmetry group of S is *transitive*; that is, elements of the symmetry group are sufficient to generate S from an arbitrary starting point.

However, in general, the symmetry group is larger than necessary to generate the signal set. For QPSK, for example, the four rotations by

0°, 90°, 180°, and 270° are sufficient to generate it. Such a subset of the symmetry group is a subgroup and, since it generates the signal set, a transitive subgroup. A transitive subgroup minimally necessary to generate the signal set is also called a *generating group*.

We see the connection to the lattices used earlier. If x_i and x_j are two lattice points in Λ, then the translation $T_{i \to j}(\Lambda) = \Lambda + (x_j - x_i)$ is an isometry that carries x_i into x_j but is invariant on the entire lattice.

Geometrically uniform signal sets need not be lattices, and Figure 3.27, taken from [28], shows the only four geometrically uniform signal sets in one dimension. Of these only set 3 is a lattice. Loosely speaking, a geometrically uniform signal set is one that "looks" identical from every signal point; that is, it does not depend on which signal point is taken as reference point; the distances to all other points are independent of this reference. In this sense it is a generalization of a lattice.

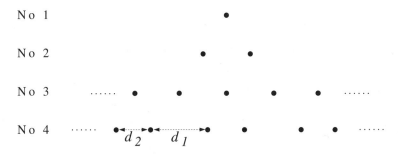

Figure 3.27 The four one-dimensional geometrically uniform signal sets.

Many signal sets used in digital communications are geometrically uniform. There are, for instance, all the lattice-based signal sets if we disregard the restriction to a finite number of signal points. Another class of geometrically uniform signal sets are the M-PSK constellations. Figure 3.28 shows an 8-PSK constellation, and it is quite obvious that rotations by multiples of $\pi/4$ leave the signal set unchanged while carrying a given signal point into any other arbitrary signal point. Also shown is an asymmetrical signal set with eight signal points that is also geometrically uniform.

If translations are involved in $T_{i \to j}(S)$, the signal set S must be infinite and periodic (e.g., signal sets 3 and 4 in Figure 3.27). If only rotations and reflections are involved, the resulting signal points lie on the surface of a sphere. Such signal sets form a *spherical code*.

As mentioned, a geometrically uniform signal set looks alike from every signal point. This statement can be made precise in a number of

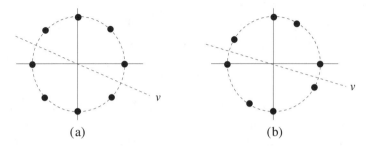

Figure 3.28 Two two-dimensional geometrically uniform constant-energy signal sets.

ways. The most interesting is that the error probability in additive Gaussian noise is the same for each signal point; that is,

$$P_e = 1 - \int_{I_{x_i}} \frac{1}{(\pi N_0)^{N/2}} \exp\left(-\frac{\|n - x_i\|^2}{N_0}\right) dn, \qquad (3.23)$$

is independent of i. This is so because the decision region (also called the *Voronoi region*) has the same shape for each signal point.

Most geometrically uniform signal sets have a generating group.[11] A generating group is obtained from the generating set of isometries of a signal set. A generating set of isometries is a minimal set of isometries that, when applied to an arbitrary initial point x_0, generates all the points of the entire constellation. For example, the set of rotations by the angles $\{0, \pi/4, \pi/2, 3\pi/4, \pi, 5\pi/4, 3\pi/2, 7\pi/4\}$, denoted by $\mathcal{R}_8$, generates the 8-PSK signal set of Figure 3.28. Alternatively the set of rotations by the angles $\{0, \pi/4, \pi/2, 3\pi/4\}$, $\mathcal{R}_4$, together with the reflection V about the axis v also generates the entire signal set; that is, we have two generating sets of isometries $\mathcal{G}_1 = \mathcal{R}_8$ and $\mathcal{G}_2 = \{V \cdot \mathcal{R}_4\}$. Note that $\mathcal{G}_2$ (but not $\mathcal{G}_1$) also generates the asymmetrical signal set in Figure 3.28. We notice that $\mathcal{G}_1$ and $\mathcal{G}_2$ have the properties of a group:

  i   They are closed under successive application of any of the isometries in the set. This is true because all isometries in the set are invariant on S; that is, $T_{i \to j}(S) = S$.

  ii They contain the unit element (rotation by 0).

  iii They contain an inverse element, which is the transformation that carries x_j back into x_i, so that the concatenation of $T_{j \to i}(x_j) \cdot$

---

[11] One would expect that all geometrically uniform signal sets have a generating group, but there exists a counterexample [29].

$T_{i \to j}(x_i) = x_i$; that is, $T_{j \to i}(x) = T_{i \to j}^{(-1)}(x)$. Such a $T_{j \to i}(x)$ must always exist since the sets we consider are generating sets; that is, they must contain a transformation that takes $x_j \to x_i$. (The initial point can be chosen as x_j.)

Therefore, from elementary group theory we know that $\mathcal{G}_1 \equiv Z_8$; that is, $\mathcal{G}_1$ is isomorphic to the additive group of integers modulo 8. We can always find an abstract group that is isomorphic to the generating set $\mathcal{G}$. We call such a group a generating group G.

We will now see how subsets of a signal set S can be associated with subgroups of its generating group G. If G' is a subgroup of G, and we will require G' to be a normal subgroup,[12] then quite obviously G' generates a subset of S. The cosets of G' in G are then associated with the different subsets S' of S. This is expressed by the following theorem, which is the major result in this section.

THEOREM 3.3

Let G' be a normal subgroup of a generating group G of S. Then the subsets of S induced by the cosets of G' in G are geometrically uniform and have G' as a common generating group (with different initial signal point x_0).

Proof

Consider the subset generated by gG', where $g \in G$. Now G' generates the subset S' to which we apply the isometry g; that is, $S_g = T_g(G')$ is congruent to S'. Now if G' is normal, $gG' = G'g$ and $S_g = \bigcup_{g' \in G'} T_{g'}\big(T_g(x_0)\big)$. Therefore, S_g is generated by G' taking the initial point $T_g(x_0)$.

Theorem 3.3 can now be used to generate geometrically uniform signal sets from a given geometrically uniform signal set by studying the normal subgroups of a generating group G. In this manner we have transformed the difficult problem of finding geometrically uniform signal sets into the simpler algebraic problem of finding normal subgroups of the generating group. For example, given an M-PSK signal set and a divisor M' of M, then $\mathcal{R}_{M'}$ is a normal subgroup of $\mathcal{R}_M$.

Or, to use our familiar lattices, if Λ' is a sublattice of Λ, the set of translations generating Λ' is isomorphic to G', which is a normal subgroup of G, the group associated with the set of translations generating Λ. For example, Z^2 is generated by all translations by (i, j), where $i, j \in \mathbf{Z}$, and $2Z^2$ is generated by all translations by $(2i, 2j)$. The original generating

---

[12] A normal subgroup is one for which the left and right cosets are identical; that is, $gG' = G'g$ for all $g \in G$.

group $G = \mathbf{Z} \times \mathbf{Z}$, and the subgroup $G' = \mathbf{Z}_e \times \mathbf{Z}_e$, where $\mathbf{Z}_e$ is the group of the even integers.

From group theory we know that a normal subgroup G' of G induces a quotient group G/G' whose elements are the cosets of the subgroup G'. Thus the elements of the quotient group G/G' are associated with the subsets S' of S; that is, there is a one-to-one map between them. If we now want to label these subsets, we may choose as labels the elements of any group isomorphic to G/G'. Let us call such a label group L, and we call a labeling with L an *isometric labeling*.

For example, the subsets of the 8-PSK signal constellation containing only one signal point can be labeled by either $\mathbf{Z}_8$, the group of integers mod 8, or by $\mathbf{Z}_2 \times \mathbf{Z}_4$, that is, by vectors (x, y) where $x \in \mathbf{Z}_2$ and $y \in \mathbf{Z}_4$. These two labelings are illustrated in Figure 3.29.

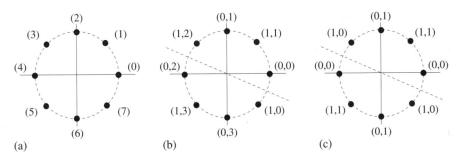

Figure 3.29 (a), (b) Two isometric labelings of the 8-PSK signal sets. The first labeling corresponds to rotations by multiples of $\pi/4$; $\mathcal{G}_a = \mathcal{R}_8$. The second labeling corresponds to reflections followed by rotations by multiples of $\pi/2$; $\mathcal{G}_b = V\mathcal{R}_4$. (c) Binary isometric labeling of the 2-PSK *subsets* generated by the subgroup $V\mathcal{R}_2$.

An isometry $T(x)$ can now be associated with an element $l \in L$; that is, applying T to a point $x(l')$ amounts to adding the labels and selecting $x(l \oplus l')$; that is, $T_l(x(l')) = x(l \oplus l')$, where $\oplus$ is the appropriate group operation.

A binary label group is a label group of the form $L = (\mathbf{Z}_2)^n$ and a labeling with the elements of L is called a *binary isometric labeling*. Binary isometric labelings are particularly interesting, since they allow the mapping of strings of bits of length n into signal points.

The importance of isometric labelings lies in the following theorem.

THEOREM 3.4
An isometric labeling of subsets is regular with respect to the label group L.

Proof

Let $l_1, l_2 \in L$ and let $x(l_1), x(l_2)$ be their corresponding signal points. Consider $d = \|x(l_1) - x(l_2)\|$. Assume that the signal point $x(0)$ is the one labeled by the unit element (0) of L. Let $T(x(l_2)) = x(0)$ be the isometry that takes $x(l_2) \rightarrow x(0)$. Since $T(x)$ is associated with $(-l_2)$, $d = \|x(l_1) - x(l_2)\| = \|T(x(l_1)) - T(x(l_2))\| = \|x(l_1 \oplus (-l_2)) - x(0)\|$ depends only on the label difference $l_1 \oplus (-l_2)$.

Earlier we defined regularity as the property that the distance between two signal points depends only on the difference between their labels, where we have used binary label groups (see (3.2)–(3.3)). Theorem 3.4 is a straightforward extension to more general label groups L.

We now understand why our search for a regular 8-PSK mapping using a binary label group $(\mathbf{Z}_2)^3$ was futile. There exists no generating group G for the 8-PSK signal set that is isomorphic to $(\mathbf{Z}_2)^3$.

However, if we look at the subsets containing two signal points generated by $\mathbf{Z}_2$ (rotation by π), the quotient group $G/G' = (V\mathcal{R}_4)/\mathcal{R}_2$ is isomorphic to $(\mathbf{Z}_2)^2$ and there exists an isometric labeling of such 2-PSK subsets, shown in Figure 3.29, where the labels are $(x, y') = (x, (y \bmod 2))$ and $L \equiv (\mathbf{Z}_2)^2$.

The labeling used originally for the 16-QAM signal set (Figure 3.10) is not regular. But using $V_v V_w T_1 T_2$, where V_v and V_w are reflections about v and w, and T_1 and T_2 are the translations whose vectors are shown in Figure 3.30, as a generating set of the coset representatives of the lattice

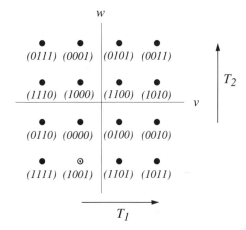

Figure 3.30 Binary isometric labeling of the cosets of $4Z^2$ in the partition $Z^2/4Z^2$. Only the coset leaders are shown, which are used as a 16-QAM constellation. The binary isometric labeling is obtained from the generating group, which is isomorphic to (V_v, V_w, T_1, T_2).

partition $Z^2/4Z^2$, an isometric labeling of the 16-QAM constellation can be found as shown in Figure 3.30.

Theorem 3.4 now allows us to identify regular trellis codes. Namely, if we combine a group trellis with an isometric labeling, such that the group of branch labels forms the label group, we have what we will call a *geometrically uniform trellis code*.

DEFINITION 3.7

A geometrically uniform trellis code using a signal constellation[13] S is generated by a group trellis, whose branch label group is an isometric labeling of the signal points in S.

From the above we immediately have the following

THEOREM 3.5

A geometrically uniform trellis code is regular; that is, the distance between any two sequences $d^2(\mathbf{x}(\mathbf{l}^{(1)}), \mathbf{x}(\mathbf{l}^{(2)})) = d^2(\mathbf{x}(\mathbf{l}^{(1)} \oplus \mathbf{l}^{(2)}), \mathbf{x}(\mathbf{0}))$, where $\mathbf{l}^{(1)}$ and $\mathbf{l}^{(2)}$ are label sequences.

The proof of this theorem follows directly from Theorem 3.4 and the additivity property of a group trellis.

While Theorem 3.5 ensures that the distance spectrum of a geometrically uniform trellis code is identical for each of the code sequences, more can be said.

THEOREM 3.6

The set of sequences (codewords) of a geometrically uniform trellis code C is geometrically uniform; that is, for any two sequences $\mathbf{x}^{(i)}, \mathbf{x}^{(j)} \in C$, there exists an isometry $T_{i \to j}$ that carries $\mathbf{x}^{(i)}$ into $\mathbf{x}^{(j)}$ and leaves the code C invariant.

Proof

From Definition 3.7 we know that the sequence $\mathbf{x}^{(i)}$ is generated by the label sequence $\mathbf{l}^{(i)}$ and that $\mathbf{x}^{(j)}$ is generated by $\mathbf{l}^{(j)}$. Since the generating trellis is a group trellis, $\mathbf{l}^{(i-j)} = \mathbf{l}^{(i)} \oplus -\mathbf{l}^{(j)}$ is also a valid label sequence. We now apply the isometry $T(x(\mathbf{l}^{(i-j)}))$ to the code sequence $\mathbf{x}^{(i)}$, in the sense that we apply $T(x(l_r^{(i-j)}))$ to $x_r^{(i)}$. By virtue of the branch label group being isomorphic to the generating group G of the signal set S used on the branches, $T(x(l_r^{(i-j)})) = T(x(l_r^{(i)} \oplus -l_r^{(j)})) = T(x(l_r^{(i)})) \cdot T^{(-1)}(x(l_r^{(i)}))$ takes signal $x_r^{(i)}$ to $x_r(0)$ and then to $x_r^{(j)}$.

---

[13] Note that this definition could be extended to allow for a different signal constellation S_r at each time interval r in a straightforward way.

Theorem 3.6 is the main theorem of this section, and it states that all code sequences in a geometrically uniform trellis code are congruent—that is, they have the same geometric structure. This includes congruent Voronoi regions, identical error performance over AWGN channels, and regularity, among other properties.

It is now a simple matter to generate geometrically uniform trellis codes. The ingredients are a group code (in our case a group trellis), and a geometrically uniform signal set S, such that the branch label group (or the group of output symbols in the general case) is a generating group of S.

Some examples of geometrically uniform trellis codes are

1. All convolutional codes are geometrically uniform.

2. Trellis codes using rate-1/2 convolutional codes and QPSK signal constellations with Gray mapping are geometrically uniform. (We need to use the reflection group V^2 to generate the QPSK set, rather than $\mathcal{R}_4$, hence Gray mapping.)

3. The four-state Ungerböck code from Table 3.1 is geometrically uniform since it uses the labeling from Figure 3.29c. None of the other 8-PSK codes are geometrically uniform. It is the job of quasi-regularity to simplify the error probability calculations.

4. The coset codes $C(\Lambda/\Lambda; C)$ of Section 3.6 are geometrically uniform with a proper binary isometric labeling of the branch label sequences C.

5. Massey et al. [30] have constructed geometrically uniform trellis codes for M-PSK signal sets using $\mathcal{R}_M$ as the label group.

3.9   HISTORICAL NOTES

Although error control coding was long regarded as a discipline with applications mostly for channel with no bandwidth limitations such as the deep-space radio channel, the trellis-coded modulation schemes of Ungerböck [1–4] provided the first strong evidence that coding could be used very effectively on band-limited channels. From the 1980s to the present, numerous researchers have discovered new results and new codes. Multidimensional schemes were first used by Wei [12] and Calderbank and Sloane [13], who also introduced the lattice viewpoint [15]. Much of the material presented in this chapter is adapted from Forney's comprehensive treatments of the subject in [26–28].

REFERENCES

[1] G. Ungerböck and I. Csajka, "On improving data-link performance by increasing channel alphabet and introducing sequence coding," *Int. Sym. Inform. Theory*, Ronneby, Sweden, June 1976.

[2] G. Ungerböck, "Channel coding with multilevel/phase signals," *IEEE Trans. Inform. Theory*, Vol. IT-28, No. 1, pp. 55–67, 1982.

[3] G. Ungerböck, "Trellis-coded modulation with redundant signal sets. I: Introduction," *IEEE Commun. Mag.*, Vol. 25, No. 2, pp. 5–11, 1987.

[4] G. Ungerböck, "Trellis-coded modulation with redundant signal sets. II: State of the art," *IEEE Commun. Mag.*, Vol. 25, No. 2, pp. 12–21, 1987.

[5] J. Du and M. Kasahara, "Improvements of the information-bit error rate of trellis code modulation systems," *IEICE, Jpn.*, Vol. E72, pp. 609–614, 1989.

[6] W. Zhang, "Finite-state machines in communications," Ph.D. thesis, University of South Australia, Australia, 1995.

[7] W. Zhang, C. Schlegel, and P. Alexander, "The BER reductions for systematic 8PSK trellis codes by a Gray scrambler," Proc. Int. Conf. on Universal Wireless Access, Melbourne, Australia, April 94.

[8] J. E. Porath and T. Aulin, "Fast algorithmic construction of mostly optimal trellis codes," Technical Report No. 5, Division of Information Theory, School of Electrical and Computer Engineering, Chalmers University of Technology, Göteborg, Sweden, 1987.

[9] J. E. Porath and T. Aulin, "Algorithmic construction of trellis codes," *IEEE Trans. Commun.*, Vol. COM-41, No. 5, pp. 649–654, 1993.

[10] J. H. Conway and N. J. A. Sloane, *Sphere Packings, Lattices and Groups*, Springer-Verlag, New York, 1988.

[11] G. D. Forney, Jr., R. G. Gallager, G. R. Lang, F. M. Longstaff, and S. U. Qureshi, "Efficient modulation for band-limited channels," *IEEE J. Select. Areas Commun.*, Vol. SAC-2, No. 5, pp. 632–647, 1984.

[12] L. F. Wei, "Trellis-coded modulation with multidimensional constellations," *IEEE Trans. Inform. Theory*, Vol. IT-33, pp. 483–501, 1987.

[13] A. R. Calderbank and N. J. A. Sloane, "An eight-dimensional trellis code," *Proc. IEEE*, Vol. 74, pp. 757–759, 1986.

[14] L. F. Wei, "Rotationally invariant trellis-coded modulations with multidimensional M-PSK,", *IEEE J. Select. Areas Commun.*, Vol. SAC-7, No. 9, pp. 1281–1295, 1989.

[15] A. R. Calderbank and N. J. A. Sloane, "New trellis codes based on lattices and cosets," *IEEE Trans. Inform. Theory*, Vol. IT-33, pp. 177–195, 1987.

[16] S. S. Pietrobon, G. U. Ungerböck, L. C. Perez, and D. J. Costello, Jr., "Rotationally invariant nonlinear trellis codes for two-dimensional modulation," *IEEE Trans. Inform. Theory*, Vol. IT-40, No. 6, pp. 1773–1791, 1994.

[17] S. S. Pietrobon, R. H. Deng, A. Lafanechère, G. Ungerböck, and D. J. Costello, Jr., "Trellis-coded multidimensional phase modulation," *IEEE Trans. Inform. Theory*, Vol. IT-36, pp. 63–89, 1990.

[18] IBM Europe, "Trellis-coded modulation schemes for use in data modems transmitting 3–7 bits per modulation interval," CCITT SG XVII Contribution COM XVII, No. D114, April 1983.

[19] L. F. Wei, "Rotationally invariant convolutional channel coding with expanded signal space. I: 180 degrees," *IEEE J. Select. Areas Commun.*, Vol. SAC-2, pp. 659–672, 1984.

[20] L. F. Wei, "Rotationally invariant convolutional channel coding with expanded signal space. II: Nonlinear codes," *IEEE J. Select. Areas Commun.*, Vol. SAC-2, pp. 672–686, 1984.

[21] IBM Europe, "Trellis-coded modulation schemes with 8-state systematic encoder and 90° symmetry for use in data modems transmitting 3–7 bits per modulation interval," CCITT SG XVII Contribution COM XVII, No. D180, October 1983.

[22] U. Black, *The V Series Recommendations, Protocols for Data Communications over the Telephone Network*, McGraw-Hill, New York, 1991.

[23] S. S. Pietrobon and D. J. Costello, Jr., "Trellis coding with multidimensional QAM signal sets," *IEEE Trans. Inform. Theory*, Vol. IT-39, pp. 325–336, 1993.

[24] M. V. Eyuboglu, G. D. Forney, P. Dong, and G. Long, "Advanced modem techniques for V.Fast," *Eur. Trans. Telecommun. ETT*, Vol. 4, No. 3, pp. 234–256, 1993.

[25] CCITT Recommendations V.34.

[26] G. D. Forney, "Coset codes. I: Introduction and geometrical classification," *IEEE Trans. Inform. Theory*, Vol. IT-34, pp. 1123–1151, 1988.

[27] G. D. Forney, "Coset codes. II: Binary lattices and related codes," *IEEE Trans. Inform. Theory*, Vol. IT-34, pp. 1152–1187, 1988.

[28] G. D. Forney, "Geometrically uniform codes," *IEEE Trans. Inform. Theory*, Vol. IT-37, pp. 1241–1260, 1991.

[29] D. Slepian, "On neighbor distances and symmetry in group codes," *IEEE Trans. Inform. Theory*, Vol. IT-17, pp. 630–632, 1971.

[30] J. L. Massey, T. Mittelholzer, T. Riedel, and M. Vollenweider, "Ring convolutional codes for phase modulation," *IEEE Int. Symp. Inform. Theory*, San Diego, CA, January 1990.

4

CONVOLUTIONAL CODES

4.1 CONVOLUTIONAL CODES AS BINARY TRELLIS CODES

In Chapter 3 we used convolutional codes as generators of our linear trellis, and the purpose of the finite-state machine was to generate the topological trellis and the linear labeling discussed in Section 3.2. This topological trellis with the linear labeling then served as the matrix for the actual trellis code. This particular way of constructing the trellis is not only of historical importance, but it also allowed us to find all the structural theorems presented in Chapter 3. Moreover, in Chapter 5, we will prove some information-theoretic results that hold for group-trellis codes. This and their widespread use make them the most important class of trellis codes.

Convolutional codes, on the other hand, have a long history. They were introduced by Elias [1] in 1955. Since then, much theory has evolved to understand convolutional codes. Some excellent references on the subject are [2–5]. In this chapter we want to bring together the main results that have evolved over the years and are elegantly summarized by Johannesson and Wan [6]. Their work mainly draws on the ground-breaking exposition by Forney [7]. Many of the details of the theory have to be sacrificed here in the interest of brevity, and our main focus will be the algebraic theory of convolutional codes.

If our trellis mapper function is omitted from the trellis code as shown in Figure 4.1, which is the finite-state machine from Figure 3.1, we have a *convolutional encoder* that produces a *convolutional code*—in other words, a code that maps a vector $u_r = (u_r^{(2)}, u_r^{(1)})$ of input bits into a vector $v_r = (v_r^{(2)}, v_r^{(1)}, v_r^{(0)})$ of output bits at time r. The rate $R = 2/3$ of this

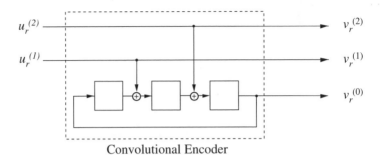

Figure 4.1 Rate $R = 2/3$ convolutional code that was used in Figure 3.1 to generate an 8-PSK trellis code.

code is the ratio of the number of input bits to the number of output bits. If these output bits are mapped individually into binary phase-shift keyed (BPSK) signals, we quickly see that a convolutional code does not conserve the signal bandwidth but requires $1/R$ times more bandwidth than uncoded transmission through the rate expansion. This is generally the case for traditional error control coding, which was mostly applied to power-limited channels such as the deep-space channel, and power efficiency is purchased at the expense of bandwidth efficiency, as discussed in Chapter 1.

Convolutional codes are traditionally used with BPSK or quadrature phase-shift keyed (QPSK) (Gray) mappings. In either case, due to the regularity of the mapper function (Section 3.3), the Euclidean distance between two signal sequences depends only on the Hamming distance $H_d(\mathbf{v}^{(1)}, \mathbf{v}^{(2)})$ between the two output bit sequences (formerly branch label sequences) $\mathbf{v}^{(1)}$ and $\mathbf{v}^{(2)}$, where the Hamming distance between two sequences is defined as the number of bit positions in which the two sequences differ. Furthermore, since the convolutional code is linear, we can choose any of the sequences as the reference sequence, and we need consider only the Hamming weights of $H_w(\mathbf{v}^{(1)} \oplus \mathbf{v}^{(2)}) = H_d(\mathbf{v}^{(1)} \oplus \mathbf{v}^{(2)}, 0) = H_d(\mathbf{v}^{(1)}, \mathbf{v}^{(2)})$, but we are getting ahead of ourselves.

The convolutional encoder in Figure 4.1 has an alternate "incarnation," which is given in Figure 4.2. This form is called the controller canonical nonsystematic form, the term stemming from the fact that inputs can be used to control the state of the encoder in a direct way, in which the outputs have no influence. We will see that both encoders generate the same code, albeit different input bit sequences map onto different output bit sequences.

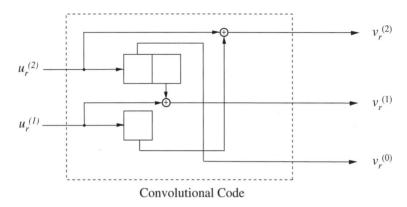

Convolutional Code

Figure 4.2 Rate $R = 2/3$ convolutional code from above in controller canonical nonsystematic form.

We see that convolutional codes are special trellis codes, and everything we said about the latter applies to the former also. The rich structure of convolutional codes, however, makes them an intriguing topic in their own right.

Figure 4.3 shows the use of convolutional codes in a communications system. First, the input sequence **u** is encoded into an output sequence **v**, which is sent over the channel. This channel contains the modulator and demodulator and it corrupts the transmitted sequence $\mathbf{v} \to \mathbf{r}$. The codeword estimator retrieves the most likely transmitted output sequence $\hat{\mathbf{v}}$ and the (pseudo) inverse of the encoder reproduces the input sequences, or at least the input sequence $\hat{\mathbf{u}}$ that corresponds to $\hat{\mathbf{v}}$.

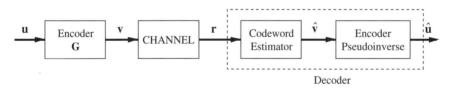

Figure 4.3 Communications system using a convolutional code with the decoder broken up into the codeword estimator and encoder inverse.

The most difficult part of the decoding operation is clearly the codeword estimator, which is a nonlinear operation. We study the codeword estimator in Chapter 5 and, particularly, in Chapter 6. This chapter is mainly concerned with the encoder/encoder inverse, since these are linear functions and a great deal can be said about them. Often the refinement of the decoder into a codeword estimator and an encoder inverse is not made in the literature and both functions are subsumed into one unit. We call the encoder inverse a pseudoinverse, since we allow for an arbitrary finite delay to take place in the output sequence $\hat{\mathbf{u}}$.

4.2   CODES AND ENCODERS

Recall (Section 3.6) the definition of the D-transform

$$u(D) = \sum_{r=s}^{\infty} u_r D^r, \tag{4.1}$$

where $s \in \mathbf{Z}$ guarantees that there are only finitely many negative terms in (4.1). Note that the D-operator can be understood as a unit-delay operator, corresponding to passing a symbol through one delay element. Formally, (4.1) is a Laurent series with vector coefficients, which we may write as two (k in general) binary Laurent series:

$$u(D) = \left(u^{(2)}(D), u^{(1)}(D)\right). \tag{4.2}$$

Note that the binary Laurent series form a field[1] and division is understood to be evaluated by expanding the long division into positive exponents, e.g., $(1 + D)/(D^2 + D^3 + D^4) = D^{-2} + 1 + D + D^3 + \cdots$.

We now have enough mathematics to formally define a convolutional code:

DEFINITION 4.1

A rate $R = k/n$ convolutional code over the field of rational Laurent series $F_2(D)$ is an injective linear mapping of the k-dimensional vector Laurent series $u(D) \in F_2^k(D)$ into the n-dimensional vector Laurent series $v(D) \in F_2^n(D)$; i.e.,

$$u(D) \rightarrow v(D) : F_2^k(D) \rightarrow F_2^n(D). \tag{4.3}$$

---

[1] That is, they possess all the field properties:

(i) Closure under addition and multiplication.

(ii) Every element $a(D)$ possesses an additive inverse $-a(D)$, and every element $a(D) \neq 0$ possesses a multiplicative inverse $1/a(D)$.

(iii) The addition operation commutes: $a(D) + b(D) = b(D) + a(D)$, and the multiplication operation commutes also: $a(D)b(D) = b(D)a(D)$.

(iv) There exists an additive unit element 0 such that $a(D) + 0 = a(D)$, and $a(D) + (-a(D)) = 0$, as well as a multiplicative unit element 1 such that $a(D) \cdot 1 = a(D)$, and $a(D)(1/a(D)) = 1$.

(v) Multiplication distributes over addition: $a(D)(b(D) + c(D)) = a(D)b(D) + a(D)c(D)$.

From basic linear algebra (see, e.g., [8]) we know that any such linear map can be represented by a matrix multiplication; in our case we write

$$v(D) = u(D)G(D),\qquad\qquad(4.4)$$

where $G(D)$ is known as a *generator matrix* of the convolutional code, or simply a *generator*, and consists of $k \times n$ entries $g_{ij}(D)$ that are Laurent series. In fact, a convolutional code is the image set of the linear operator $G(D)$.

We will concentrate on *delay-free* generator matrices, that is, those that have no common multiple of D in the numerator of $g_{ij}(D)$. In other words, a general generator matrix $G_n(D)$ can always be written as

$$G_n(D) = D^i G(D),\qquad i \geq 1,\qquad\qquad(4.5)$$

by pulling out the common term D^i of all $g_{ij}(D)$, where i is the delay. Note that this restriction does not affect the generality of the results in this chapter.

From Definition 4.1 we see that a convolutional code is the set of output sequences, irrespective of the particular mapping of input-to-output sequences. There exist an infinite number of generator matrices for the same code, and we define the equivalence of two generator matrices in

DEFINITION 4.2

Two generator matrices $G(D)$ and $G'(D)$ are equivalent, written $G(D) \equiv G'(D)$, if they generate the same convolutional code $\{v(D)\}$, where $\{v(D)\}$ is the set of all possible output sequences $v(D)$.

Examining Figure 4.2, we can read off quite easily that

$$G_2(D) = \begin{bmatrix} 1 & D^2 & D \\ D & 1 & 0 \end{bmatrix}\qquad\qquad(4.6)$$

and, upon further examination,[2] that

$$v(D) = u(D)G_2(D) = u(D)\begin{bmatrix} 1 & D^2 \\ D & 1 \end{bmatrix}\begin{bmatrix} 1 & 0 & \frac{D}{1+D^3} \\ 0 & 1 & \frac{D^2}{1+D^3} \end{bmatrix}$$

$$= u'(D)\begin{bmatrix} 1 & 0 & \frac{D}{1+D^3} \\ 0 & 1 & \frac{D^2}{1+D^3} \end{bmatrix} = u'(D)G_1(D),\qquad\qquad(4.7)$$

---

[2] Note that all coefficient operations are in the field GF(2), that is, XOR operations.

where $G_1(D)$ is the generator matrix for the encoder in Figure 4.1. The set of sequences $\{u(D)\}$ is identical to the set of sequences $\{u'(D)\}$ if and only if

$$T(D) = \begin{bmatrix} 1 & D^2 \\ D & 1 \end{bmatrix} \Longrightarrow T^{-1}(D) = \frac{1}{1+D^3}\begin{bmatrix} 1 & D^2 \\ D & 1 \end{bmatrix} \qquad (4.8)$$

is invertible and $u'(D) = u(D)T(D)$.

Since we have cast our coding world into the language of linear systems, the matrix $T(D)$ is invertible if and only if its determinant $\det(T(D)) = 1 + D^3 \neq 0$. In this case the set of possible input sequences $\{u(D)\}$ is mapped onto itself; $\{u(D)\} = \{u'(D)\}$, since $T(D)$ is a $k \times k$ matrix of full rank.

The generator $G_1(D)$ is called *systematic*, corresponding to the systematic encoder[3] of Figure 4.1, since the input bits $(u_r^{(2)}, u_r^{(1)})$ appear unaltered as $(v_r^{(2)}, v_r^{(1)})$. Both $G_1(D)$ and $G_2(D)$ have the same number of states, as can be seen from Figures 4.1 and 4.2, and they both generate the same code $\{v(D)\}$.

Appealing to linear systems theory again, we know that there are an infinite number of matrix representations for a given linear map, each such representation amounting to a different choice of bases. Another such choice, or generator, for our code is

$$G_3(D) = \begin{bmatrix} 1+D & 1+D^2 & D \\ D+D^2 & 1+D & 0 \end{bmatrix}. \qquad (4.9)$$

This new encoder has four delay elements in its realization and therefore has more states than $G_1(D)$ or $G_2(D)$. But there is a more serious problem with $G_3(D)$. Since we can always transform one basis representation into another basis representation via a matrix multiplication, we find that

$$G_3(D) = \begin{bmatrix} 1 & 1 \\ 0 & 1+D \end{bmatrix} G_2(D), \qquad (4.10)$$

and the input sequence

$$u(D) = \left[0, \frac{1}{1+D} = 1 + D + D^2 + \cdots \right] \qquad (4.11)$$

generates the output sequence

$$v(D) = \left[0, \frac{1}{1+D}\right] G_3(D) = [D, 1, 0], \qquad (4.12)$$

---

[3] We will use the terms generator matrix and encoder interchangeably, realizing that while $G(D)$ may have different physical implementations, these differences are irrelevant from our viewpoint.

whose Hamming weight $H_w(v(D)) = 2$. We have the unsettling case that an infinite-weight ($H_w(u(D)) = \infty$) input sequence generates a finite-weight output sequence. Such an encoder is called *catastrophic*, since a finite number of channel errors in the reception of $v(D)$ can cause an infinite number of errors in the data $u(D)$. Such encoders are to be avoided. We define formally:

DEFINITION 4.3

An encoder $G(D)$ for a convolutional code is catastrophic if there exists a $u(D)$ such that $H_w(u(D)) = \infty$ and $H_w(u(D)G(D)) < \infty$.

Note that the property of being catastrophic is one of the encoder, because although $G_3(D)$ is catastrophic, neither $G_2(D)$ nor $G_1(D)$ are, and all generate the same code! We now note that the problem stemmed from the fact that

$$T_3^{-1}(D) = \begin{bmatrix} 1 & 1 \\ 0 & 1+D \end{bmatrix}^{-1} = \begin{bmatrix} 1 & \frac{1}{1+D} \\ 0 & \frac{1}{1+D} \end{bmatrix} \tag{4.13}$$

was not well behaved, which turned $G_1(D)$ into a catastrophic $G_3(D)$. The problem was that some of the entries of $T_3^{-1}(D)$ have an infinite number of coefficients; i.e., they are fractional.

Since for a catastrophic encoder, a finite-weight sequence $v(D)$ maps into an infinite-weight sequence $u(D)$, the encoder right inverse[4] $G^{-1}(D)$ must have fractional entries. We therefore require that for a "useful" encoder, $G^{-1}(D)$ must have no fractional entries—actually all its entries are required to be polynomials in D; that is, they have no negative powers and only a finite number of nonzero coefficients. This then is a sufficient condition for $G(D)$ not to be catastrophic. We will see that it is also a necessary condition. Note that both $G_1(D)$ and $G_2(D)$ have polynomial right inverses:

$$G_1^{-1}(D) = \begin{bmatrix} 1 & 0 \\ 0 & 1 \\ 0 & 0 \end{bmatrix} \tag{4.14}$$

and

$$G_2^{-1}(D) = \begin{bmatrix} 1 & D \\ D & 1+D^2 \\ D^2 & 1+D+D^3 \end{bmatrix} \tag{4.15}$$

and are therefore not catastrophic.

---

[4] Such an inverse must always exist, since we defined a convolutional code to be an *injective* map, i.e., one that can be inverted.

Let us further define the class of basic encoders as follows:

DEFINITION 4.4

An encoder $G(D)$ is basic if it is polynomial and has a polynomial right inverse.

$G_2(D)$, for example, is a basic encoder. We will use basic encoders as a main tool to develop the algebraic theory of convolutional codes. For a basic encoder we define the constraint length

$$\nu = \sum_{i=1}^{k} \max_{j} \left(\deg \left(g_{ij}(D) \right) \right) = \sum_{i=1}^{k} \nu_i. \tag{4.16}$$

Note that $\nu_i = \max_j \left(\deg \left(g_{ij}(D) \right) \right)$, the maximum degree among the polynomials in row i of $G(D)$, corresponds to the maximum number of delay units that the ith input bits $u^{(i)}(D)$ need to be stored in the encoder in the controller canonical realization.

Again, we find for $G_2(D)$ that $\nu = 3$; the number of delay elements needed in the controller canonical encoder realization (Figure 4.2). Since the inputs are binary, the number of states of $G_2(D)$ is $S = 2^\nu = 8$, and we see that for basic encoders the states and the number of states are easily defined and determined.

One wonders whether 2^ν is the minimum number of states necessary to generate a certain convolutional code. To pursue this question further we need the following

DEFINITION 4.5

A minimal basic encoder is a basic encoder that has the smallest constraint length among all equivalent basic encoders.

We will see that $G_2(D)$ is indeed minimal.

As another example of an equivalent encoder consider

$$G_4(D) = T(D)G_2(D) = \begin{bmatrix} 1 & 1+D+D^2 \\ 0 & 1 \end{bmatrix} \begin{bmatrix} 1 & D^2 & D \\ D & 1 & 0 \end{bmatrix} \tag{4.17}$$

$$= \begin{bmatrix} 1+D+D^2+D^3 & 1+D & D \\ D & 1 & 0 \end{bmatrix}, \tag{4.18}$$

whose constraint length $\nu = 4$. Note that, since

$$\begin{bmatrix} 1 & 1+D+D^2 \\ 0 & 1 \end{bmatrix}^{-1} = \begin{bmatrix} 1 & 1+D+D^2 \\ 0 & 1 \end{bmatrix}, \tag{4.19}$$

$G_4(D)$ has a polynomial inverse

$$G_4^{-1}(D) = G_2^{-1}(D) \begin{bmatrix} 1 & 1+D+D^2 \\ 0 & 1 \end{bmatrix} \qquad (4.20)$$

$$= \begin{bmatrix} 1 & D \\ D & 1+D^2 \\ D^2 & 1+D+D^3 \end{bmatrix} \begin{bmatrix} 1 & 1+D+D^2 \\ 0 & 1 \end{bmatrix} \qquad (4.21)$$

$$= \begin{bmatrix} 1 & 1+D^2 \\ D & 1+D+D^3 \\ D^2 & 1+D+D^2+D^4 \end{bmatrix}. \qquad (4.22)$$

We have seen that two encoders are equivalent if and only if $T(D)$ has a nonzero determinant and is therefore invertible. We may now strengthen this result for basic encoders.

THEOREM 4.1
Two basic encoders $G(D)$ and $G'(D)$ are equivalent if and only if $G(D) = T(D)G'(D)$, where $T(D)$ is a $k \times k$ polynomial matrix with unit determinant, $\det(T(D)) = 1$.

PROOF
Since both $T(D)$ and $G'(D)$ are polynomial, and since $T(D)$ has full rank, $\{u(D)G(D)\} = \{u(D)T(D)G'(D)\} = \{u'(D)G'(D)\}$, and $G(D) \equiv G'(D)$.

Conversely, if $G(D) \equiv G'(D)$, $T^{-1}(D) = \mathrm{adj}(T(D))/\det(T(D))$ must exist.[5] Since $G(D)$ is basic, it has a polynomial right inverse, $G^{-1}(D)$, and $G'(D)G^{-1}(D) = T^{-1}(D)$ is polynomial also. Therefore $G'(D) = T^{-1}(D)T(D)G'(D)$ and $T^{-1}(D)T(D) = I_k$, the $k \times k$ identity matrix. But since both $T(D)$ and $T^{-1}(D)$ are polynomial, and $\det(I_k) = 1 = \det(T(D))\det(T^{-1}(D))$, we conclude that $\det(T(D)) = 1$.

The binary polynomials in D, denoted by $F[D]$, form a commutative ring; i.e., they possess all the field properties except division. Other "famous" examples of rings are the integers, $\mathbf{Z}$, and the integers modulo m, $\mathbf{Z}_m$. Certain elements in a ring do have inverses: they are called *units*. In $\mathbf{Z}$ the units are $\{-1, 1\}$, whereas in $F[D]$ it is only the unit element 1. In the proof of Theorem 4.1, we could also have used the following basic algebraic result [8, page 96].

THEOREM 4.2
A square matrix G with elements from a commutative ring R is invertible if and only if $\det(G) = r_u$, where $r_u \in R$ is a unit—in other words, if and only if $\det(G)$ is invertible in R.

---

[5] The cofactor $T_{ij}(D)$ of the element $t_{ij}(D)$ in $T(D)$ is $(-1)^{i+j}$ times the determinant of the $k-1 \times k-1$ matrix obtained by striking out the ith row and the jth column of $T(D)$. The adjoint $\mathrm{adj}(T(D))$ is the $k \times k$ matrix whose (i, j)th entry is $T_{ji}(D)$ (note the reversal in the order of the subscripts).

A square polynomial matrix $T(D)$ with a polynomial inverse is also called a *scrambler*, since such a $T(D)$ will simply scramble the input sequences $\{u(D)\}$—that is, relabel them.

4.3   FUNDAMENTAL THEOREMS FROM BASIC ALGEBRA

With the preliminaries from the last section, we are now ready for our first major theorem. Let us decompose the basic encoder $G(D)$ into two parts,

$$G(D) = \tilde{G}(D) + \hat{G}(D) = \tilde{G}(D) + \begin{bmatrix} D^{\nu_1} & & & \\ & D^{\nu_2} & & \\ & & \ddots & \\ & & & D^{\nu_k} \end{bmatrix} G_h, \qquad (4.23)$$

where G_h is a matrix with $(0, 1)$ entries, a 1 indicating the position where the highest-degree term D_i^ν occurs in row i; for example,

$$G_2(D) = \begin{bmatrix} 1 & 0 & D \\ 0 & 1 & 0 \end{bmatrix} + \begin{bmatrix} D^2 & \\ & D \end{bmatrix} \begin{bmatrix} 0 & 1 & 0 \\ 1 & 0 & 0 \end{bmatrix} \qquad (4.24)$$

and

$$G_4(D) = \begin{bmatrix} 1 + D + D^2 & 1 + D & D \\ 0 & 1 & 0 \end{bmatrix} + \begin{bmatrix} D^3 & \\ & D \end{bmatrix} \begin{bmatrix} 1 & 0 & 0 \\ 1 & 0 & 0 \end{bmatrix}. \qquad (4.25)$$

A minimal basic encoder is then characterized by [6]

THEOREM 4.3
A polynomial $G(D)$ is a minimal basic encoder if and only if

   i G_h has full rank $(\det(G_h) \neq 0)$ or, equivalently, if and only if

   ii   the maximum degree of all $k \times k$ subdeterminants of $G(D)$ equals the constraint length ν.

PROOF
Assume that G_h does not have full rank. Then there exists a sum of $d \leq k$ rows $\mathbf{h}_{i_j}$ of G_h such that

$$\mathbf{h}_{i_1} + \mathbf{h}_{i_2} + \cdots + \mathbf{h}_{i_d} = 0. \qquad (4.26)$$

Assume now that we have ordered the indices in decreasing maximum row degrees, such that $\nu_{i_d} \geq \nu_{i_j}, d \geq j$. Adding

$$D^{\nu_{i_d}} \left(\mathbf{h}_{i_1} + \mathbf{h}_{i_2} + \cdots + \mathbf{h}_{i_{d-1}} \right) \qquad (4.27)$$

to row i_d of $\hat{G}(D)$ reduces it to an all-zero row; similarly, adding

$$D^{v_{i_d}-v_{i_1}}\mathbf{g}_{i_1} + D^{v_{i_d}-v_{i_2}}\mathbf{g}_{i_2} + \cdots + D^{v_{i_d}-v_{i_{d-1}}}\mathbf{g}_{i_{d-1}} \tag{4.28}$$

to row i_d of $G(D)$ will reduce the highest degree of row i_d and produce an equivalent generator. The new generator matrix now has a constraint length less than that of the original generator matrix, and we have a contradiction to the original assumption that our $G(D)$ was minimal basic.

After some thought (working with $\tilde{G}(D)$), it is quite easy to see that conditions (i) and (ii) in the theorem are equivalent.

Since Theorem 4.3 has a constructive proof, there follows a simple algorithm to obtain a minimal basic encoder from any basic encoder, given by

Step 1. If G_h has full rank, $G(D)$ is a minimal basic encoder and we stop, else

Step 2. Let $\mathbf{h}_{i_j}$, $j \leq k$, be a set of rows of G_h such that

$$\mathbf{h}_{i_1} + \mathbf{h}_{i_2} + \cdots + \mathbf{h}_{i_d} = 0. \tag{4.29}$$

Further, let $\mathbf{g}_{i_j}$ be the corresponding rows of $G(D)$ and add

$$D^{v_{i_d}-v_{i_1}}\mathbf{g}_{i_1} + D^{v_{i_d}-v_{i_2}}\mathbf{g}_{i_2} + \cdots + D^{v_{i_d}-v_{i_{d-1}}}\mathbf{g}_{i_{d-1}} \tag{4.30}$$

to row i_d of $G(D)$. Go to Step 1.

Note that a minimal basic encoder for a given convolutional code is not necessarily unique. For example,

$$G_5(D) = \begin{bmatrix} 1 & D \\ 0 & 1 \end{bmatrix}\begin{bmatrix} 1 & D^2 & D \\ D & 1 & 0 \end{bmatrix} \tag{4.31}$$

$$= \begin{bmatrix} 1+D^2 & D^2 & D \\ D & 1 & 0 \end{bmatrix}, \tag{4.32}$$

is a basic encoder with constraint length $v = 3$ equivalent to $G_2(D)$, and both are minimal.

Our treatment of convolutional encoders has focused on basic encoders so far. We now justify why this class of encoders is so important. But before we can continue, we need some more basic algebra.

Figure 4.4 relates the algebraic concepts needed in the rest of this chapter. We start with the (commutative) ring R. If R is commutative and has no nonzero zero divisors (i.e., there exist no nonzero numbers $r_1, r_2 \in R$ such that $r_1 r_2 = 0$), the ring R is called an *integral domain*. In an integral domain we may use the cancelation law: $r_1 r_2 = r_1 r_3 \Rightarrow r_2 = r_3$, which

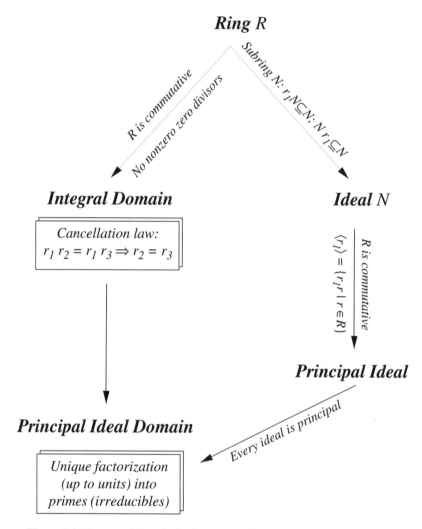

Figure 4.4   Diagram of ring-algebraic concepts. Some examples of rings to which these concepts apply are the integer numbers **Z**, the integers modulo m, **Z**$_m$ and the polynomials in D over the field F, $F[D]$.

is something like division but not quite as powerful. Examples of integral domains are **Z**, $F[D]$ (the ring of binary polynomials), and **Z**$_p$, the integers modulo p when p is a prime.

On the other hand, a subring N of a ring R is an *ideal* if, for every $r \in R$, $rN \in N$ and $Nr \in N$. That is, the ideal brings every element of the original ring R into itself through multiplication. If the ring R (and hence also N) is commutative, we define a principal ideal N as one generated by all the multiples of a single element n with R; that is, $\langle n \rangle = \{rn | r \in R\}$. An example of a principal ideal in $\mathbf{Z}$ is $\langle 2 \rangle$, the ideal of all even numbers. Now, if every ideal in a commutative ring R is principal, we have a *principal ideal domain*. This is an algebraic structure with powerful properties, one of the most well known of which is

THEOREM 4.4
UNIQUE FACTORIZATION THEOREM In a principal ideal domain, every element can be factored uniquely, up to unit elements, into primes or irreducibles. These are elements that cannot be factored.

In $\mathbf{Z}$ this is the popular integer prime factorization. Remember that the units in $\mathbf{Z}$ are 1 and -1, so the primes can be taken positive or negative by multiplying by -1. One usually agrees on the convention to take only positive primes. Technically this factorization applies to fields also (such as $\mathbf{Z}_p$), but, alas, in a field every element is a unit, so the factorization is not unique. In $F[D]$, however, the only unit is 1 and we have an unambiguous factorization.

For principal ideal domains we have the important (see, e.g., [8, page 181ff. and [7])

THEOREM 4.5
INVARIANT FACTOR THEOREM Given a $k \times n$ matrix $\mathbf{P}$ with elements from the principal ideal domain R, $\mathbf{P}$ can be written as

$$\mathbf{P} = \mathbf{A}\Gamma\mathbf{B}, \tag{4.33}$$

where

$$\Gamma = \begin{bmatrix} \gamma_1 & & & & & \\ & \gamma_2 & & & 0 & \\ & & \ddots & & & \\ & & & \gamma_k & & \\ & & & & 0 & \\ & 0 & & & & \ddots \\ & & & & & & 0 \end{bmatrix}, \tag{4.34}$$

and $\gamma_i | \gamma_j$, if $i \leq j$. Furthermore, the $k \times k$ matrix $\mathbf{A}$ and the $n \times n$ matrix $\mathbf{B}$ have unit determinants and are therefore invertible in R. The invariant factors are $\gamma_i = \Delta_i / \Delta_{i-1}$, where Δ_i is the greatest common divisor (g.c.d.) of the $i \times i$ subdeterminants of $\mathbf{P}$.

The invariant factor theorem can be extended to rational matrices. Let $\mathbf{R}$ be a $k \times n$ matrix whose entries are fractions of elements of R. Furthermore, let ϕ be the least common multiple of all denominators of the entries in $\mathbf{R}$. We may now apply the invariant factor theorem to the matrix $\mathbf{P} = \phi \mathbf{R}$, which has all its elements in R, and obtain

$$\phi \mathbf{R} = \mathbf{A}\mathbf{\Gamma}'\mathbf{B} \Rightarrow \mathbf{R} = \mathbf{A}\mathbf{\Gamma}\mathbf{B}, \tag{4.35}$$

where the new entries of Γ are $\gamma_i = \gamma_i'/\phi$, where, again, $\gamma_i | \gamma_j$ for $i \le j$, since $\gamma_i' | \gamma_j'$.

Applying these concepts to our description of encoders, we find the following decomposition for a generating matrix $G(D)$ with rational entries:

$$G(D) = A(D)\Gamma(D)B(D), \tag{4.36}$$

where $A(D)$ is a $k \times k$ scrambler and $B(D)$ is an $n \times n$ scrambler. Now, since the last $n - k$ diagonal elements of $\Gamma(D) = 0$, we may strike out the last $n - k$ rows in $B(D)$ and obtain a $k \times n$ polynomial matrix, which we call $G_b(D)$. Since $A(D)$ and $\Gamma(D)$ are invertible, $G(D) \equiv G_b(D)$; that is, $G(D) = A(D)\Gamma(D)G_b(D)$ and $G_b(D)$ generate the same code. But $G_b(D)$ has entries in $F[D]$, and, furthermore, since $B(D)$ has a unit determinant and a polynomial inverse, so does $G_b(D)$. We conclude that $G_b(D)$ is a basic encoding matrix that is equivalent to the original rational encoding matrix $G(D)$. Hence the following

THEOREM 4.6
Every rational encoding matrix has an equivalent basic encoding matrix.

In this discussion we have also developed the following algorithm to construct an equivalent basic encoding matrix for any rational encoding matrix:

Step 1.   Compute the invariant factor decomposition of the rational matrix $G(D)$, given by

$$G(D) = A(D)\Gamma(D)B(D). \tag{4.37}$$

Step 2.   Delete the last $n - k$ rows of $B(D)$ to obtain the equivalent basic $k \times n$ encoding matrix $G_b(D)$.

An algorithm to calculate the invariant factor decomposition is presented in Appendix 4.A.

We now have a pretty good idea about basic encoders and how to obtain a minimal version thereof, but the question of whether some nonpolynomial encoders might have a less complex internal structure is still daunting. We know how to turn any encoder into a basic encoder and then into a minimal basic encoder, but do we lose anything in doing so? We are now ready to address this question, and we will see that a minimal basic encoder is minimal in a more general sense than we have been able to show up to now. In order to do so, we need the concept of abstract states.

An abstract state is, loosely speaking, the internal state of an encoder that produces a particular output ($v(D)$ in our case) if the input sequence is all zero. This is known as the zero-input response. Different abstract states must generate different output sequences; otherwise they are not distinguishable and are the same state. Obviously different abstract states correspond to different physical states (the states of the shift registers in an encoder implementation), but different physical states might correspond to the same abstract state. This will appear again in Chapter 5 as state equivalence. Hence, the number of abstract states will always be equal to or smaller than the number of physical states of a given encoder. The number of abstract states is therefore a more basic measure of the inherent complexity of an encoder.

To generate all possible abstract states, we simply use all possible input sequences from $-\infty$ to time unit 0, at which time we turn off the input sequences and observe the set of possible outputs $\{v(D)\}$. This set we define as our abstract states. To formalize, let P be the projection operator that sets the input sequence to 0 for all nonnegative time units— that is, $u(D)P = (\ldots, \mathbf{u}_{-2}, \mathbf{u}_{-1}, 0, 0, 0, \ldots)$—and let Q be the projection operator that sets the output sequence to zero for all negative time units— that is, $v(D)Q = (\ldots, 0, 0, \mathbf{v}_0, \mathbf{v}_1, \ldots)$. The abstract state corresponding to an input sequence $u(D)$ is then formally given by

$$v_S(D) = (u(D)PG(D))\, Q. \tag{4.38}$$

Our first result is

THEOREM 4.7
The number of abstract states of a minimal basic encoder is 2^ν, equal to the number of physical states.

Proof
Let

$$s(D) = \left(u^{(k)}_{-\nu_k} D^{-\nu_k} + \cdots + u^{(k)}_{-1} D^{-1}, \ldots, u^{(1)}_{-\nu_1} D^{-\nu_1} + \cdots + u^{(1)}_{-1} D^{-1} \right)$$

$$\tag{4.39}$$

be a physical state of the basic encoder in its controller canonical form. For example, for the code from Figure 4.2, $S(D) = (u_{-2}^{(2)}D^{-2} + u_{-1}^{(2)}D^{-1}, u_{-1}^{(1)}D^{-1})$. There are 2^ν different physical states in our basic encoder. Let us now assume that two different physical states, $s_1(D)$ and $s_2(D)$, correspond to the same abstract state:

$$v(D)Q = s_1(D)G(D)Q = s_2(D)G(D)Q. \tag{4.40}$$

This is equivalent to

$$(s_1(D) - s_2(D)) \, G(D)Q = s(D)G(D)Q = 0; \tag{4.41}$$

that is, there exists a nonzero state $s(D)$ whose corresponding abstract state is zero, according to (4.41). We now show that the only state that fulfills (4.41) is $s(D) = 0$; hence $s_1(D) = s_2(D)$. This will prove the theorem by contradiction.

We need to have $v(D) = 0$ (i.e., $v_i = 0$ for all $i \geq 0$) for (4.41) to hold. Now, without loss of generality, assume that

$$\nu_k = \nu_{k-1} = \cdots = \nu_{k-l} > \nu_{k-l-1} \geq \cdots \geq \nu_1. \tag{4.42}$$

The coefficient $v_{\nu_{k-1}}$ of D^{ν_k-1} in $v(D)$ is

$$v_{\nu_{k-1}} = \left(0, \ldots, 0, u_{-1}^{(k-l)}, \ldots, u_{-1}^{(k)}, \right) G_h(D) = 0, \tag{4.43}$$

but, since $G_h(D)$ has full rank for a minimal basic encoding matrix, we must have $u_{-1}^{(k)} = u_{-1}^{(k-1)} = \cdots = u_{-1}^{(k-\ell)} = 0$. Continuing by induction, we then prove $u_{-2}^{(k)} = u_{-2}^{k-1} = \cdots = u_{-2}^{(k-\ell-1)} = 0$, etc., that is, that $s(D) = 0$, which proves the theorem.

We see that the abstract states capture the memory of an encoder $G(D)$ in a very general way, and we are therefore interested in finding the $G(D)$ that minimizes the number of abstract states, realizing that this is all we need to keep track of in the decoder. Hence, we make the following

DEFINITION 4.6

A minimal encoder is an encoder $G(D)$ that has the smallest number of abstract states over all equivalent encoders (basic or not).

If $G(D)$ and $G'(D)$ are two equivalent encoders with abstract states $v_S(D)$ and $v_S'(D)$, respectively, we can relate these abstract states as follows:

$$\begin{aligned} v_S(D) &= u(D)PG(D)Q = u(D)PT(D)G'(D)Q \\ &= u(D)PT(D)(P + Q)G'(D)Q \\ &= u(D)PT(D)PG'(D)Q + u(D)PT(D)QG'(D)Q, \end{aligned} \tag{4.44}$$

but the first term is an abstract state of $G'(D)$, denoted $v'_S(D)$, and the second term is a codeword:

$$v'(D) = u(D)PT(D)QG'(D)Q \tag{4.45}$$

$$= u(D)PT(D)QG'(D), \tag{4.46}$$

where we were allowed to drop the operator Q in equation (4.46) since for any *realizable* encoder $G'(D)$ has to be a causal encoding matrix. Therefore, since $u'(D) = u(D)PT(D)Q$ is an input sequence that starts at time 0 (i.e., $u'_j = 0$, $j < 0$) and since $G'(D)$ must be causal, $v'(D)$ cannot have any nonzero negative components either. Hence, the trailing Q-operator in (4.46) is superfluous, and we have the representation given in

THEOREM 4.8

$$v_S(D) = v'_S(D) + v'(D), \tag{4.47}$$

where, if $G'(D)$ is a minimal basic encoder, the representation (4.47) is unique.

Proof
To see this, assume the contrary, namely

$$v_{S_{(mb)}}(D) + v(D) = v'_{S_{(mb)}}(D) + v'(D); \tag{4.48}$$

that is,

$$v_{S_{(mb)}}(D) + v'_{S_{(mb)}}(D) = v''_{S_{(mb)}}(D) = v''(D) = v(D) + v'(D), \tag{4.49}$$

and $v''(D)$ is both a codeword and an abstract state $v''_{S_{(mb)}}(D)$ of the minimal basic encoder. But then

$$v''(D) = u''(D)G'(D) \Rightarrow u''(D) = v''(D)[G'(D)]^{-1}(D), \tag{4.50}$$

and $u''(D)$ is polynomial since $[G'(D)]^{-1}$ is polynomial. Furthermore, since $v''_{S_{(mb)}}(D) = u(D)G'(D)Q$, for some input sequence $u(D)$ with no nonzero terms u_j, $j \geq 0$, and

$$v''_{S_{(mb)}}(D) = u(D)G'(D)Q = u''(D)G'(D), \tag{4.51}$$

we obtain

$$\bigl(u''(D) + u(D)\bigr)G'(D)Q = 0. \tag{4.52}$$

Using the same method as in the proof of Theorem 4.7, one can show that $\bigl(u''(D) + u(D)\bigr) = 0$, which implies $u''(D) = u(D) = 0$ and $v''_{S_{(mb)}} = 0$, leading to a contradiction. This proves the theorem.

Due to $v_{S_{(mb)}}(D) = v_S(D) + v(D)$ and the uniqueness of (4.47), the map $v_S(D) \to v_{S_{(mb)}}(D) : v_S(D) = v_{S_{(mb)}}(D) + v'(D)$ is surjective (i.e., onto), and we have our next theorem.

THEOREM 4.9
The number of abstract states of an encoder $G(D)$ is always larger than or equal to the number of abstract states of an equivalent minimum basic encoder $G_{mb}(D)$.

We also immediately notice the following

COROLLARY 4.10
A minimal basic encoder is a minimal encoder.

This proves that the minimal basic encoders are desirable since they represent an implementation with the minimum number of abstract states, yet more can be said about minimal encoders in general, expressed by the following central

THEOREM 4.11
$G(D)$ is a minimal encoder if and only if

i   its number of abstract states equals the number of abstract states of an equivalent minimal basic encoder or if and only if

ii $G(D)$ has a polynomial right inverse in D and a polynomial right inverse in D^{-1}.

PROOF

i   Part (i) is obvious from Theorem 4.9.

ii Let $u(D)$ be given and assume that $v(D) = u(D)G(D)$ is polynomial in D^{-1}; i.e., $v_r = 0$ for $r > 0$ and $r < s$, where $s \leq 0$ is arbitrary. Then

$$D^{-1}v(D)Q = 0. \tag{4.53}$$

Breaking this term into two components, we get

$$
\begin{aligned}
D^{-1}v(D)Q &= D^{-1}u(D)(P + Q)G(D)Q \\
&= D^{-1}u(D)PG(D)Q + D^{-1}u(D)QG(D)Q. \tag{4.54}
\end{aligned}
$$

Since $G(D)$ is causal, $D^{-1}u(D)QG(D)Q = D^{-1}u(D)QG(D)$ is a codeword. Therefore, the abstract state $D^{-1}u(D)PG(D)Q$ is a codeword, so, following (4.41)ff., together with Theorem 4.8, we conclude that it must be the zero codeword. Since $G(D)$ has full rank, $D^{-1}u(D)QG(D) = 0$ implies

$$D^{-1}u(D)Q = 0; \tag{4.55}$$

that is, $u(D)$ contains no positive powers of D. Since $v(D) = u(D)G(D)$ and $G(D)$ is delay free, $u(D)$ must be polynomial in D^{-1}; i.e., $u_r = 0$ for $r < s$. Therefore the map $v(D) \to u(D) = G^{-1}(D)v(D)$ maps polynomials in D^{-1} into polynomials in D^{-1}, and $G^{-1}(D)$, the right inverse of $G(D)$, must be polynomial in D^{-1} also.

The fact that a minimal encoder $G(D)$ also has a polynomial right inverse in D is proven similarly. Assume now that $v(D) = u(D)G(D)$ is polynomial in D. Then

$$v(D) = u(D)PG(D) + u(D)QG(D), \qquad (4.56)$$

where $u(D)Q$ is a power series (only positive terms). Then, due to causality, $u(D)QG(D)$ is also a power series, which implies that $u(D)QG(D)$ must also be a power series. But now

$$u(D)PG(D) = u(D)PG(D)Q, \qquad (4.57)$$

is again a codeword and an abstract state, which implies that it must be the zero codeword; that is, $u(D)PG(D) = 0$. But since $G(D)$ has full rank we conclude that $u(D)P = 0$; that is, $u(D)$ is a power series. Now from above take $G^{-1}(D^{-1})$, a polynomial right inverse in D^{-1}. Then $G^{-1}(D) = D^s G^{-1}(D^{-1})$ is a (pseudo)inverse that is polynomial in D, such that

$$u(D)D^s = v(D)G^{-1}(D^{-1}), \qquad (4.58)$$

and hence a polynomial $v(D)$ can only generate a polynomial $u(D)$; that is, $u(D)$ has only finitely many terms. Therefore the inverse map $G^{-1}(D)\colon v(D) \to u(D) = v(D)G^{-1}(D)$ must also be polynomial.

Now for the reverse part of the theorem assume that $G(D)$ has a polynomial inverse in D^{-1} and in D. Assume further that the abstract state $v_s(D) = u(D)G(D)Q = u'(D)G(D)$ is a codeword and that $u(D)$ is polynomial in D^{-1} without a constant term. $v_s(D)$ is therefore a power series, and since $G(D)$ has a polynomial inverse in D by assumption, it follows that $u'(D)$ is also a power series. Now, using the right inverse $G^{-1}(D^{-1})$, which is polynomial in D^{-1}, we write

$$u'(D)G(D)G^{-1}(D^{-1}) = u(D)G(D)QG^{-1}(D^{-1})$$
$$u'(D) = u(D)G(D)QG^{-1}(D^{-1}). \qquad (4.59)$$

But the left side in (4.59) has no negative powers in D, and the right side has no positive powers since $u(D)G(D)Q$ has no positive powers and $G^{-1}(D^{-1})$ is polynomial in D^{-1}. We conclude that $v_s(D) = 0$. Using Theorems 4.8 and 4.9, we conclude that $G(D)$ is minimal.

As pointed out by Johannesson and Wan, part (ii) of the theorem provides us with a practical minimality test for encoders. Furthermore, since a minimal encoder $G(D)$ has a polynomial right inverse $(G(D))^{-1}$, it follows immediately that $G(D)$ is not catastrophic. Hence,

COROLLARY 4.12

A minimal encoder is not catastrophic.

Johannesson and Wan [6] pointed out that there are minimal encoders that are not minimal basic. Quoting their example, the basic encoder

$$G(D) = \begin{bmatrix} 1 + D & D & 1 \\ 1 + D^2 + D^3 & 1 + D + D^2 + D^3 & 0 \end{bmatrix} \qquad (4.60)$$

has constraint length $\nu = 4$ but is not minimal basic. In fact

$$G'(D) = \begin{bmatrix} 1 + D & D & 1 \\ D^2 & 1 & 1 + D + D^2 \end{bmatrix} \qquad (4.61)$$

is an equivalent minimal basic encoder with constraint length $\nu' = 3$.

Nevertheless, $G(D)$ has a polynomial right inverse in D^{-1},

$$\left(G(D^{-1})\right)^{-1} = \begin{bmatrix} 1 + D^{-1} + D^{-2} + D^{-3} & D^{-1} \\ 1 + D^{-1} + D^{-3} & D^{-1} \\ D^{-2} + D^{-3} & D^{-1} \end{bmatrix}, \qquad (4.62)$$

and is therefore, according to Theorem 4.11, a minimal encoder with $2^3 = 8$ abstract states.

4.4   SYSTEMATIC ENCODERS

The encoder in Figure 4.1, whose encoding matrix is

$$G_s(D) = \begin{bmatrix} 1 & 0 & \frac{D}{1+D^3} \\ 0 & 1 & \frac{D^2}{1+D^3} \end{bmatrix}, \qquad (4.63)$$

is systematic; that is, the k information bits appear unchanged in the output sequence $v(D)$.

We now show that every code has a systematic encoder. Let us assume without loss of generality that the code is generated by the basic encoder $G(D)$. Then $G^{-1}(D)$ exists and is a polynomial matrix. Using the invariant factor theorem we write

$$G^{-1}(D) = \left(A(D)\Gamma(D)B(D)\right)^{-1} = B^{-1}(D)\Gamma^{-1}(D)A^{-1}(D). \qquad (4.64)$$

Since $B^{-1}(D)$ and $A^{-1}(D)$ are polynomial, $\Gamma^{-1}(D)$ must be polynomial also. This means that all the invariant factors must be units (i.e., $\gamma_i = 1$), and thus the greatest common divisor (g.c.d.) of the $k \times k$ subdeterminants of $G(D)$ must equal 1. It follows that there exists a $k \times k$ subdeterminant $\Delta_k(D)$ of $G(D)$, which is a delay-free polynomial (no multiple of D), since otherwise the g.c.d. of all the $k \times k$ subdeterminant would be a multiple of D.

We now rearrange the columns of $G(D)$ such that the first k columns form that matrix $T(D)$ whose determinant is $\Delta_k(D)$. Since $T(D)$ is invertible we form

$$G_s(D) = T^{-1}(D)G(D) = [\mathbf{I}_k | P(D)], \tag{4.65}$$

where $P(D)$ is a $n - k \times k$ parity-check matrix with (possibly) rational entries. For example, for the code from Figure 4.1

$$P(D) = \begin{bmatrix} \dfrac{D}{1 + D^3} \\ \dfrac{D^2}{1 + D^3} \end{bmatrix}. \tag{4.66}$$

Note that the systematic encoder for a given code is unique, whereas there may exist several equivalent minimal basic encoders for the same code.

Systematic encoders have another nice property, given by

THEOREM 4.13
Every systematic encoder for a convolutional code is minimal.

Proof
The proof of this result comes easily from Theorem 4.11. Consider the inverse of a systematic encoder, which is simply the $k \times k$ identity matrix, which, trivially, is both polynomial in D and in D^{-1}.

To obtain a systematic encoder for a given code, the procedure outlined earlier is practical; that is, we find a $k \times k$ submatrix whose determinant is a polynomial that is not a multiple of D. Then we simply calculate (4.65) to obtain $G_s(D)$.

The inverse operation is more difficult. Given a systematic encoder $G_s(D)$, we want to find an equivalent minimal basic encoder. We can proceed by calculating the invariant factor decomposition of $G_s(D)$. This will give us a basic encoder. Then we apply the algorithm on page 101 to obtain a minimal basic encoder.

Often it is easier to apply a more ad hoc technique, based on examining the trellis of a given code [11]. Let us take the example of Figure 4.1 (equation (4.63)). This systematic encoder has eight states. Since systematic encoders are minimal, we are looking for a minimal basic encoder with eight states. The code rate is 2/3, and therefore the two values for v_1, v_2 must be 1 and 2 (or 0 and 3, which are quickly ruled out). The minimal basic encoder blueprint for our code is shown in Figure 4.5.

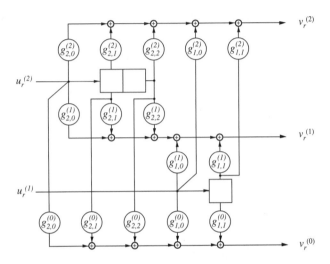

Figure 4.5 Encoder blueprint for the minimal basic encoder equivalent to $G_S(D)$.

We now look at the trellis generated by the systematic encoder, whose first few transitions originating from the zero state are shown in Figure 4.6. We start by tracing paths from the zero state back to the zero state and map them into the connectors $g_{1,m}^{(j)}$ and $g_{2,m}^{(j)}$. The first merger occurs after two branches with the output sequence $(v_0, v_1) = ((010), (100))$. The first triple v_0 is generated by setting $g_{1,0}^{(2)} = 0$, $g_{1,0}^{(1)} = 1$, and $g_{1,0}^{(0)} = 0$. The second triple v_1 is generated by setting $g_{1,1}^{(2)} = 1$, $g_{1,1}^{(1)} = 0$, and $g_{1,1}^{(0)} = 0$. Now we move onto the next path. It is advantageous to chose paths for which $u_r^{(1)} = 0$, since then the connectors $g_{1,m}^{(j)}$ will not interfere. One such path is (100), (001), (010). From this we can determine the connectors $g_{2,0}^{(2)} = 1$, $g_{2,0}^{(1)} = 0$, and $g_{2,0}^{(0)} = 0$. From the second triple we obtain $g_{2,1}^{(2)} = 0$, $g_{2,1}^{(1)} = 0$, and $g_{2,1}^{(0)} = 1$ and, from the third triple $g_{2,2}^{(2)} = 0$, $g_{2,2}^{(1)} = 1$, and $g_{2,2}^{(0)} = 0$. Checking back with Figure 4.2, we see that we have obtained the same encoder. (Note that choosing other paths could have generated another, equivalent, encoder.) This procedure can easily be generalized for larger encoders.

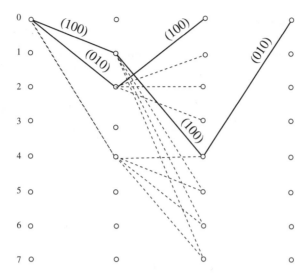

Figure 4.6 Initial trellis section generated by the systematic encoder from Figure 4.1.

4.5   MAXIMUM FREE-DISTANCE CONVOLUTIONAL CODES

If the output bits $v(D)$ of a convolutional code are mapped into the signals $\{-1, +1\}$ of a BPSK signal constellation, the minimum squared Euclidean distance d_{free}^2 depends only on the number of binary differences between the closest code sequences. This number is the minimum Hamming distance of a convolutional code, denoted by d_{free}. Since convolutional codes are linear, finding the minimum Hamming distance between two sequences $v^{(1)}(D)$ and $v^{(2)}(D)$, $H_d(v^{(1)}(D), v^{(2)}(D))$, amounts to finding the minimum Hamming weight of any code sequence $v(D)$. Finding convolutional codes with large minimum Hamming distance is as difficult as finding good trellis codes with large Euclidean free distance, and, as in the case for trellis codes, computer searches are usually used to find good codes [12–14]. Most often the controller canonical form (Figure 4.2) of an encoder is preferred in these searches; that is, the search is targeted at finding a minimal basic encoder. The procedure is then to search for a code with the largest minimum Hamming weight by varying the connector taps $g_{i,m}^{(j)}$ either exhaustively or according to heuristic rules. Once an encoder is found it is tested for minimality, which will then ensure that it is not catastrophic.

In this fashion the codes in Tables 4.1–4.8 were found [3, 13, 15, 16]. They are the rate $R = 1/2$, $R = 1/3$, and $R = 2/3$ codes with the largest minimum Hamming distance d_{free} for a given constraint length.

TABLE 4.1 Connectors and Free Hamming Distance of the Best $R = 1/2$ Convolutional Codes [3]

Constraint length ν	$g^{(1)}$	$g^{(0)}$	d_{free}
2	5	7	5
3	15	17	6
4	23	35	7
5	65	57	8
6	133	171	10
7	345	237	10
8	561	753	12
9	1161	1545	12
10	2335	3661	14
11	4335	5723	15
12	10533	17661	16
13	21675	27123	16
14	56721	61713	18
15	111653	145665	19
16	347241	246277	20

Source: The connectors are given in octal notation, e.g., $g = 17 = 111$.

TABLE 4.2 Connectors and Free Hamming Distance of the Best $R = 1/3$ Convolutional Codes [3]

Constraint length ν	$g^{(2)}$	$g^{(1)}$	$g^{(0)}$	d_{free}
2	5	7	7	8
3	13	15	17	10
4	25	33	37	12
5	47	53	75	13
6	133	145	175	15
7	225	331	367	16
8	557	663	711	18
9	1117	1365	1633	20
10	2353	2671	3175	22
11	4767	5723	6265	24
12	10533	10675	17661	24
13	21645	35661	37133	26

TABLE 4.3   Connectors and Free Hamming Distance of the Best $R = 2/3$ Convolutional Codes [3]

Constraint length v	$g_2^{(2)}, g_1^{(2)}$	$g_2^{(1)}, g_1^{(1)}$	$g_2^{(0)}, g_1^{(0)}$	d_{free}
2	3, 1	1, 2	1, 2	3
3	2, 1	1, 4	1, 4	4
4	7, 2	1, 5	1, 5	5
5	14, 3	6, 10	6, 10	6
6	15, 6	6,15	6,15	7
7	14, 3	7, 11	7, 11	8
8	32, 13	5, 33	5, 33	8
9	25, 5	3, 70	3, 70	9
10	63, 32	15, 65	15, 65	10

TABLE 4.4   Connectors and Free Hamming Distance of the Best $R = 1/4$ Convolutional Codes [13]

Constraint length v	$g^{(3)}$	$g^{(2)}$	$g^{(1)}$	$g^{(0)}$	d_{free}
2	5	7	7	7	10
3	13	15	15	17	13
4	25	27	33	37	16
5	53	67	71	75	18
6	135	135	147	163	20
7	235	275	313	357	22
8	463	535	733	745	24
9	1117	1365	1633	1653	27
10	2387	2353	2671	3175	29
11	4767	5723	6265	7455	32
12	11145	12477	15537	16727	33
13	21113	23175	35527	35537	36

TABLE 4.5   Connectors and Free Hamming Distance of the Best $R = 1/5$ Convolutional Codes [16]

Constraint length v	$g^{(4)}$	$g^{(3)}$	$g^{(2)}$	$g^{(1)}$	$g^{(0)}$	d_{free}
2	7	7	7	5	5	13
3	17	17	13	15	15	16
4	37	27	33	25	35	20
5	75	71	73	65	57	22
6	175	131	135	135	147	25
7	257	233	323	271	357	28

The field of convolutional codes is naturally much larger than has been covered in this summary chapter. For further treatment of convolutional codes see Lin and Costello [3] and Dholakia [2].

TABLE 4.6   Connectors and Free Hamming Distance of the Best $R = 1/6$ Convolutional Codes [16]

Constraint length v	$g^{(5)}$	$g^{(4)}$	$g^{(3)}$	$g^{(2)}$	$g^{(1)}$	$g^{(0)}$	d_{free}
2	7	7	7	7	5	5	16
3	17	17	13	13	15	15	20
4	37	35	27	33	25	35	24
5	75	75	55	65	47	57	27
6	173	151	135	135	163	137	30
7	253	375	235	235	313	357	34

TABLE 4.7   Connectors and Free Hamming Distance of the Best $R = 1/7$ Convolutional Codes [16]

Constraint length v	$g^{(6)}$	$g^{(5)}$	$g^{(4)}$	$g^{(3)}$	$g^{(2)}$	$g^{(1)}$	$g^{(0)}$	d_{free}
2	7	7	7	7	5	5	5	18
3	17	17	13	13	13	15	15	23
4	35	27	25	27	33	35	37	28
5	53	75	65	75	47	67	57	32
6	165	145	173	135	135	147	137	36
7	275	253	375	331	235	313	357	40

TABLE 4.8   Connectors and Free Hamming Distance of the Best $R = 1/8$ Convolutional Codes [16]

Constraint length v	$g^{(7)}$	$g^{(6)}$	$g^{(5)}$	$g^{(4)}$	$g^{(3)}$	$g^{(2)}$	$g^{(1)}$	$g^{(0)}$	d_{free}
2	7	7	5	5	5	7	7	7	21
3	17	17	13	13	13	15	15	17	16
4	37	33	25	25	35	33	27	37	32
5	57	73	51	65	75	47	67	57	36
6	153	111	165	173	135	135	147	137	40
7	275	275	253	371	331	235	313	357	45

Appendix

We will generate a sequence of matrix operations that turn **P** into the diagonal matrix Γ according to Theorem 4.5 (see also [8, Section 3.7] or [7]). First, we define some elementary matrix operations of size $k \times k$ or $n \times n$, depending on whether we premultiply or postmultiply **P**. The first such operation is $\mathbf{T}_{ij}(b) = \mathbf{I} + b\mathbf{E}_{ij}$, where **I** is the identity matrix, $\mathbf{E}_{ij}$ is a matrix with a single 1 in position (i, j) and zeros elsewhere, and $b \in R$, the principal ideal domain.[6] $\mathbf{T}_{ij}(b)$ is invertible in R, since

$$\mathbf{T}_{ij}(b)\mathbf{T}_{ij}(-b) = \left(\mathbf{I} + b\mathbf{E}_{ij}\right)\left(\mathbf{I} + (-b)\mathbf{E}_{ij}\right) = \mathbf{I}. \tag{4.67}$$

Left multiplication of $\mathbf{P}$ by a $k \times k$ matrix $\mathbf{T}_{ij}(b)$ adds b times the jth row to the ith row of $\mathbf{P}$, leaving the remaining rows unchanged. Right multiplication of $\mathbf{P}$ by an $n \times n$ matrix $\mathbf{T}_{ij}(b)$ adds b times the ith column to the jth column of $\mathbf{P}$.

Next, let u be a unit in R and define $\mathbf{D}_i(u) = \mathbf{I} + (u-1)\mathbf{E}_{ii}$. Again, $\mathbf{D}_i(u)$ is invertible with inverse $\mathbf{D}_i(-u)$. Left multiplication with $\mathbf{D}_i(u)$ multiplies the ith row of $\mathbf{P}$ by u, while right multiplication multiplies the ith column by u.

Finally, define $\mathbf{Q}_{ij} = \mathbf{I} - \mathbf{E}_{ii} - \mathbf{E}_{jj} + \mathbf{E}_{ij} + \mathbf{E}_{ji}$, and $\mathbf{Q}_{ij}$ is its own inverse. Left multiplication by $\mathbf{Q}_{ij}$ interchanges the ith and jth rows of $\mathbf{P}$, while right multiplication interchanges the ith and jth rows, leaving the remainder of the matrix unchanged.

Last, define

$$\mathbf{U} = \begin{bmatrix} x & s & & & & \\ y & t & & & 0 & \\ & & 1 & & & \\ & & & \ddots & & \\ & 0 & & & 1 & \\ & & & & & 1 \end{bmatrix}, \tag{4.68}$$

where the submatrix $\begin{bmatrix} x & s \\ y & t \end{bmatrix}$ is invertible.

Let us start now, and assume $p_{11} \neq 0$ (otherwise bring a nonzero element into its position via elementary operations) and that p_{11} does not divide p_{12} (otherwise, again, move such an element into its position via elementary operations). Now, according to Euclid's division theorem [8], there exist elements $x, y \in R$ such that

$$p_{11}x + p_{12}y = \gcd(p_{11}, p_{12}) = d, \tag{4.69}$$

where d, the greatest common divisor of p_{11} and p_{12}, can be found via Euclid's algorithm. Let $s = p_{12}/d$ and $t = -p_{11}/d$ in (4.68), which makes it invertible:

$$\begin{bmatrix} x & p_{12}/d \\ y & -p_{11}/d \end{bmatrix}^{-1} = \begin{bmatrix} p_{11}/d & p_{12}/d \\ y & -x \end{bmatrix}. \tag{4.70}$$

---

[6] It might be helpful to think in terms of integers; that is, $R = \mathbf{Z}$.

Multiplying **P** on the right by this matrix gives a matrix whose first row is $(d, 0, p_{13}, \ldots, p_{1k})$. If d does not divide all elements of the new matrix, move such an element into position $(2, 1)$ via elementary matrix operations and repeat the process until p'_{11} divides all p'_{ij} for all i, j.

Via elementary matrix operations the entire first row and column can be cleared to zero, leaving the equivalent matrix

$$\mathbf{P}'' = \begin{bmatrix} \gamma_1 & 0 \\ 0 & \mathbf{P}''_1 \end{bmatrix}, \tag{4.71}$$

where $\gamma_1 = p''_{11}$, up to units, and γ_1 divides every element in $\mathbf{P}''_1$, since the elementary matrix operations do not affect divisibility.

Iterating this procedure now, continuing with $\mathbf{P}''_1$ yields

$$\mathbf{P}''' = \begin{bmatrix} \gamma_1 & 0 & 0 \\ 0 & \gamma_2 & 0 \\ 0 & 0 & \mathbf{P}'''_2 \end{bmatrix}, \tag{4.72}$$

where the divisibility $\gamma_1 | \gamma_2$ is ensured since γ_1 divides all elements in $\mathbf{P}''_1$, which implies that any elementary matrix operation preserves this divisibility, and γ_2 divides every element in $\mathbf{P}'''_2$.

We now simply iterate this procedure to obtain the decomposition in Theorem 4.5, in the process producing the desired matrices **A** and **B**. It can further be shown that this decomposition is unique in the sense that all equivalent matrices yield the same diagonal matrix Γ (see [8, Section 3.7]).

Note that this procedure also provides a proof for Theorem 4.5.

REFERENCES

[1] P. Elias, "Coding for noisy channels," *IRE Conv. Rec.,* Pt. 4, pp. 37–47, 1955.

[2] A. Dholakia, *Introduction to Convolutional Codes,* Kluwer, Boston, 1994.

[3] S. Lin and Daniel J. Costello, Jr., *Error Control Coding: Fundamentals and Applications,* Prentice Hall, Englewood Cliffs, NJ, 1983.

[4] G. C. Clark and J. B. Cain, *Error-Correction Coding for Digital Communications,* Plenum Press, New York, 1983.

[5] A. J. Viterbi and J. K. Omura, *Principles of Digital Communication and Coding,* McGraw-Hill, New York, 1979.

[6] R. Johannesson and Z.-X. Wan, "A linear algebra approach to minimal convolutional encoders," *IEEE Trans. Inform. Theory,* Vol. IT-39, No. 4, pp. 1219–1233, 1993.

[7] G. D. Forney, "Convolutional codes. I: Algebraic structure," *IEEE Trans. Inform. Theory*, Vol. IT-16, No. 6, pp. 720–738, 1970.

[8] N. Jacobson, *Basic Algebra I*, Freeman, New York, 1985.

[9] J. L. Massey and M. K. Sain, "Inverses of linear sequential circuits," *IEEE Trans. Computers*, Vol. C-17, pp. 310–337, April 1968.

[10] M. K. Sain and J. L. Massey, "Invertibility of linear time-invariant dynamical systems," *IEEE Trans. Automat. Control,* Vol. AC-14, pp. 141–149, April 1969.

[11] J. E. Porath, "Algorithms for converting convolutional codes from feedback to feedforward form and vice versa," *Electron. Lett.*, Vol. 25, No. 15, pp. 1008–1009, 1989.

[12] J. P. Odenwalder, "Optimal decoding of convolutional codes," Ph.D. thesis, University of California, Los Angeles, 1970.

[13] K. J. Larsen, "Short convolutional codes with maximum free distance for rates 1/2, 1/3, and 1/4," *IEEE Trans. Inform. Theory*, Vol. IT-19, pp. 371–372, May 1973.

[14] E. Paaske, "Short binary convolutional codes with maximum free distance for rates 2/3 and 3/4," *IEEE Trans. Inform. Theory*, Vol. IT-20, pp. 683–689, 1974.

[15] J. G. Proakis, *Digital Communications*, 3rd ed., McGraw-Hill, New York, 1995.

[16] D. G. Daut, J. W. Modestino, and L. D. Wismer, "New short constraint length convolutional code construction for selected rational rates," *IEEE Trans. Inform. Theory*, Vol. IT-28, pp. 793–799, 1982.

CHAPTER 5

PERFORMANCE BOUNDS

5.1 THE ERROR EVENT PROBABILITY

In this chapter we will concern ourselves with the performance evaluation of trellis codes. We have seen that the encoder for a trellis code is a finite-state machine (FSM) whose transitions are determined by the input data sequence. The optimal trellis decoder is a copy of this FSM that attempts to retrace the path (i.e., the state sequence) taken by the encoder FSM (see Chapter 6). The decoder will make an error if the path it follows through its trellis does not coincide with the one taken by the encoder. Such a scenario is illustrated in Figure 5.1. Here the decoder starts an error at time interval j by following an erroneous path and remerges with the correct sequence at time interval $j + L$.

An important measure of interest is the information bit error probability (BER) associated with our decoder. Unfortunately, it turns out that the calculation of BER is extremely complex and no efficient methods exist. We therefore look for more accessible ways of obtaining a measure of performance. We proceed in the same way as usual for block codes and consider the probability that a codeword error occurs or, more appropriately, for trellis codes a code sequence error occurs. Such an error happens when the decoder follows a path that diverges from the correct path somewhere in the trellis.

Figure 5.2 shows an example trellis where the correct path is indicated by a solid line, and all possible paths (i.e., all error paths) by dotted lines. The total length of the trellis and the code is l, and we assume that the code is always started in the zero state and terminated in the zero state.

The probability of error, P, is then the probability that any of the dotted paths in Figure 5.2 is chosen; that is, P is the probability of the union of

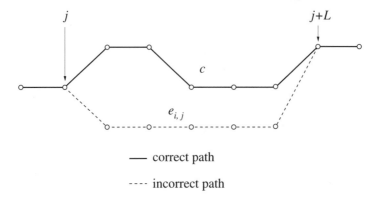

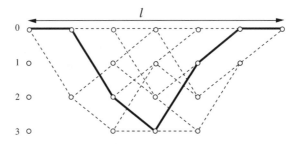

Figure 5.1 The correct path and an error path of length L in a trellis diagram.

Figure 5.2 Trellis diagram showing a given correct path and the set of possible error paths.

the individual errors $e_{i,j}$, given by

$$P = \Pr\left(\bigcup_j \bigcup_i e_{i,j} \middle| c\right), \tag{5.1}$$

where $e_{i,j}$ is the ith error path departing from the correct path c at time unit j. To obtain an average error probability we also need to average over all correct paths:

$$\overline{P} = \sum_c p(c)\Pr\left(\bigcup_j \bigcup_i e_{i,j} \middle| c\right), \tag{5.2}$$

where $p(c)$ is the probability that the encoder chooses path c.

The probability in (5.2) is still difficult to calculate and we use the union bound[1] to simplify the expression further to obtain

---

[1] The union bound states that the probability of the union of distinct events is smaller or equal to the sum of the probabilities of the individual events; i.e., $\Pr(\bigcup E_i) \leq \sum_i \Pr(E_i)$.

$$\overline{P} \le \sum_c p(c) \sum_j \Pr\left(\bigcup_i e_{i,j} \Big| c\right). \tag{5.3}$$

If the total length l of the encoded sequence is very large, $\overline{P}$ will be very large; in fact, it will approach 1 as $l \to \infty$, and its probability is therefore not a good measure for the performance of a trellis code. We therefore normalize $\overline{P}$ per time unit and consider

$$\overline{P_e} = \lim_{l \to \infty} \frac{1}{l}\overline{P}. \tag{5.4}$$

Since an infinite trellis looks identical at every time unit we can eliminate the sum over j in (5.3) to obtain

$$\overline{P_e} \le \sum_c p(c)\Pr\left(\bigcup_i e_i \Big| c\right), \tag{5.5}$$

where e_i is the event than an error starts at an arbitrary but fixed time unit, say j. Also, the correct sequence c upto node j and after node $j + L$ is irrelevant for the error path e_i of length L. The right side of (5.5) can also be interpreted as the *first event error probability*—that is, the probability that the decoder starts its first error event at node j. $\overline{P_e}$ is bounded above by the first event error probability.

To make the bound manageable, we apply the union bound again to obtain

$$\overline{P_e} \le \sum_c p(c) \sum_{e_i} \Pr\left(e_i \big| c\right). \tag{5.6}$$

Let us denote $\Pr\left(e_i \big| c\right)$ by $P_{c \to e_i}$, which, since there are only two hypotheses involved, is easily evaluated as (see equation (2.14))

$$P_{c \to e_i} = Q\left(\sqrt{d_{ci}^2 \frac{RE_b}{2N_0}}\right), \tag{5.7}$$

where $R = k/n$ is the code rate in bits/symbol, N_0 is the one-sided noise power spectral density, E_b is the energy per information bit, and d_{ci}^2 is the normalized squared Euclidean distance between the signals on the error path e_i and the signals on the correct path c. The normalization is such that the signal constellation used has unit average energy. Equation (5.7) is commonly called the *pairwise error probability*.

The upper bound on $\overline{P_e}$ now becomes

$$\overline{P_e} \le \sum_c p(c) \sum_{e_i \mid c} Q\left(\sqrt{d_{ci}^2 \frac{RE_b}{2N_0}}\right), \tag{5.8}$$

which can be rearranged as

$$\overline{P_e} \leq \sum_{\substack{i \\ (d_i^2 \in \mathcal{D})}} A_{d_i^2} Q \left(\sqrt{d_i^2 \frac{RE_b}{2N_0}} \right),$$ (5.9)

by counting how often each of the distances d_i^2 occurs in (5.8). $\mathcal{D}$ is the set of all possible squared Euclidean distances d_i^2 that occur between the signals on c and e_i in a particular trellis code, and $A_{d_i^2}$ is the average number of times d_i^2 occurs, termed the *average multiplicity* of d_i^2. The smallest d_i^2 that can be found in the trellis is called d_{free}^2, the free squared Euclidean distance, or *free distance* for short.

The infinite set of pairs $\{d^2, A_{d^2}\}$ is called the *distance spectrum* of the code, and we immediately see its connection to the error probability of the code. Figure 5.3 shows the distance spectrum of the 16-state 8-PSK code from Table 3.1, whose free squared Euclidean distance $d_{\text{free}}^2 = 5.17$.

From the averaged error event probability we can obtain a bound on the average bit error probability (BER) by the following reasoning. Each error

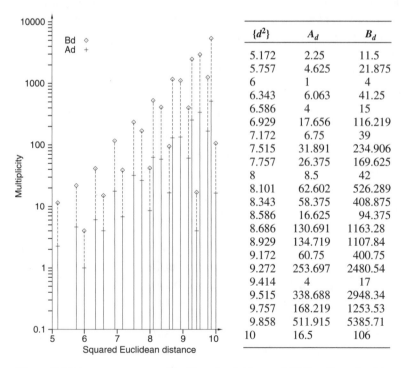

$\{d^2\}$	A_d	B_d
5.172	2.25	11.5
5.757	4.625	21.875
6	1	4
6.343	6.063	41.25
6.586	4	15
6.929	17.656	116.219
7.172	6.75	39
7.515	31.891	234.906
7.757	26.375	169.625
8	8.5	42
8.101	62.602	526.289
8.343	58.375	408.875
8.586	16.625	94.375
8.686	130.691	1163.28
8.929	134.719	1107.84
9.172	60.75	400.75
9.272	253.697	2480.54
9.414	4	17
9.515	338.688	2948.34
9.757	168.219	1253.53
9.858	511.915	5385.71
10	16.5	106

Figure 5.3 Distance spectrum of the Ungerböck 16-state 8-PSK trellis code with generators $h^{(0)} = 23$, $h^{(1)} = 4$, and $h^{(2)} = 16$.

event $c \to e_i$ will cause a certain number of bit errors. If we replace A_{d^2} with B_{d^2}, which is the averaged number of bit errors on error paths with distance d^2, we obtain a bound on the bit errors per time unit. Since our trellis code processes k bits per time unit, the average bit error probability is bounded by

$$\overline{P_b} \leq \sum_{\substack{i \\ (d_i^2 \in \mathcal{D})}} \frac{1}{k} B_{d_i^2} Q\left(\sqrt{d_i^2 \frac{RE_b}{2N_0}}\right). \tag{5.10}$$

Figure 5.4 shows the bound (5.10) for some popular 8-PSK Ungerböck trellis codes (solid curves) and compares it to results obtained through simulations (dashed curves).

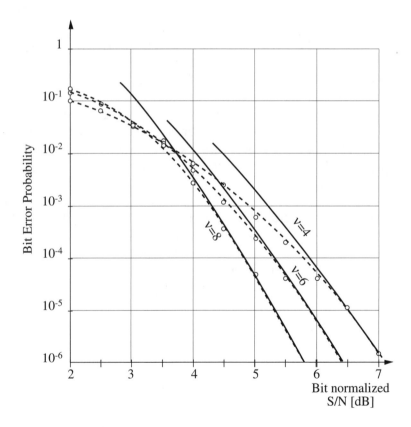

Figure 5.4 Bound on the bit error probability and simulation results for the three 8-PSK Ungerböck trellis codes with $\nu = 4, 6, 8$ from Table 3.1. The simulation results are from [1].

5.2   A FINITE-STATE MACHINE DESCRIPTION OF THE ERROR EVENTS

If a trellis code is regular (or geometrically uniform), the averaging over c in (5.8) is not necessary and any code sequence may be taken as reference sequence. For geometrically uniform codes, the Voronoi regions of all the code sequences are congruent, and hence the error probability is the same for all code sequences. For the more general case of regular codes, the distances between sequences depend only on their label differences, and the distance spectrum is identical for all sequences. Hence, for the purpose of distance spectrum calculations via (5.8), the state reduction method discussed in this section is not needed. The code trellis itself can be used for searching the distance spectrum (with respect to a chosen reference). However, many trellis codes are not regular, and we wish to have available a more general method.

We know that both the encoder and the optimal decoder are identical finite-state machines M. Since we are interested in the set of squared Euclidean distances that can occur if those two FSMs follow different paths, we find it helpful to consider a new FSM $\mathcal{M} = M \otimes M$ with states (p, q), $p, q \in M$, and outputs $\delta((p, q) \rightarrow (p_1, q_1)) = d^2_{(p,q),(p_1,q_1)}$, where $d^2_{(p,q),(p_1,q_1)}$ is the distance increment accrued when the correct path c advances from state p to p_1 and the incorrect path e_i advances from q to q_1. If the transition $(p, q) \rightarrow (p_1, q_1)$ is not possible, $\delta((p, q) \rightarrow (p_1, q_1)) = \infty$, by definition.

Let us pause here and discuss what motivates the construction of $\mathcal{M}$. Consider the case where the correct path progresses through the states $p \rightarrow p_1 \rightarrow p_2 \rightarrow \cdots \rightarrow p_{L-1} \rightarrow p_L$ and the error path leads through the states $q \rightarrow q_1 \rightarrow q_2 \rightarrow \cdots \rightarrow q_{L-1} \rightarrow q_L$, where $q = p$ and $q_L = p_L$. This is illustrated in Figure 5.5 for $L = 5$.

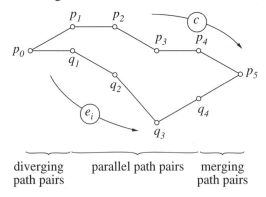

diverging   parallel path pairs   merging
path pairs   path pairs

Figure 5.5 The correct path and an error path of length $L = 5$.

If we now multiply the exponentials of the outputs $\delta((p_{i-1}, q_{i-1}) \to (p_i, q_i))$ of $\mathcal{M}$ we obtain

$$\prod_{i=1}^{L} X^{\delta((p_{i-1}, q_{i-1}) \to (p_i, q_i))} = X^{\sum_{i=1}^{L} d^2_{(p_{i-1}, q_{i-1}), (p_i, q_i)}} = X^{d^2_{ci}}; \qquad (5.11)$$

that is, the total squared Euclidean distance between c and e_i appears in the exponent of our dummy base X.

Note that $\mathcal{M}$ is a random FSM; that is, we are not assigning any inputs to the transitions. Each transition is taken with equal probability $1/2^k$, which is the probability of the transition $p \to p_1$ of the correct path c.

We can now define the output transition matrix associated with our FSM $\mathcal{M}$, whose entries are given by

$$\mathbf{B} = \{b_{(pq)(p_1 q_1)}\} = \left\{ \frac{1}{2^k} X^{\delta((p,q) \to (p_1, q_1))} \right\} = \left\{ \frac{1}{2^k} X^{d^2_{(p,q), (p_1, q_1)}} \right\}, \qquad (5.12)$$

and it is quite straightforward to see that the $((p, p), (q, q))$-entry of the Lth power of $\mathbf{B}$ is a polynomial in X whose exponents are all the distances between path pairs originating in (p, p) and terminating in (q, q), and whose coefficients are the average multiplicities of these distances. The weighting factor $1/2^k$ weighs each transition with the probability that c moves from $p \to p_1$. This achieves the weighting of the distances with the probability of c. We now can use matrix multiplication to keep track of the distances between paths.

$$\mathbf{B} = \frac{1}{4} \begin{array}{c} \\ 00 \\ 01 \\ 02 \\ 03 \\ 10 \\ 11 \\ 12 \\ 13 \\ 20 \\ 21 \\ 22 \\ 23 \\ 30 \\ 31 \\ 32 \\ 33 \end{array} \begin{array}{cccccccccccccccc} 00 & 01 & 02 & 03 & 10 & 11 & 12 & 13 & 20 & 21 & 22 & 23 & 30 & 31 & 32 & 33 \\ \left(\begin{array}{cccccccccccccccc} 1 & X^2 & X^4 & X^2 & X^2 & 1 & X^2 & X^4 & X^4 & X^2 & 1 & X^2 & X^2 & X^4 & X^2 & 1 \\ X^{3.4} & X^{0.6} & X^{0.6} & X^{3.4} & X^{3.4} & X^{3.4} & X^{0.6} & X^{0.6} & X^{0.6} & X^{3.4} & X^{3.4} & X^{0.6} & X^{0.6} & X^{0.6} & X^{3.4} & X^{3.4} \\ X^2 & 1 & X^2 & X^4 & 1 & X^2 & X^4 & X^2 & X^2 & X^4 & X^2 & 1 & X^4 & X^2 & 1 & X^2 \\ X^{0.6} & X^{3.4} & X^{3.4} & X^{0.6} & X^{3.4} & X^{3.4} & X^{0.6} & X^{0.6} & X^{3.4} & X^{0.6} & X^{0.6} & X^{3.4} & X^{0.6} & X^{0.6} & X^{3.4} & X^{3.4} \\ X^{3.4} & X^{3.4} & X^{0.6} & X^{0.6} & X^{0.6} & X^{3.4} & X^{3.4} & X^{0.6} & X^{0.6} & X^{0.6} & X^{3.4} & X^{3.4} & X^{3.4} & X^{0.6} & X^{0.6} & X^{3.4} \\ 1 & X^2 & X^4 & X^2 & X^2 & 1 & X^2 & X^4 & X^4 & X^2 & 1 & X^2 & X^2 & X^4 & X^2 & 1 \\ X^{3.4} & X^{3.4} & X^{0.6} & X^{0.6} & X^{3.4} & X^{0.6} & X^{0.6} & X^{3.4} & X^{0.6} & X^{0.6} & X^{3.4} & X^{3.4} & X^{0.6} & X^{3.4} & X^{3.4} & X^{0.6} \\ X^2 & 1 & X^2 & X^4 & 1 & X^2 & X^4 & X^2 & X^2 & X^4 & X^2 & 1 & X^4 & X^2 & 1 & X^2 \\ X^2 & 1 & X^2 & X^4 & 1 & X^2 & X^4 & X^2 & X^2 & X^4 & X^2 & 1 & X^4 & X^2 & 1 & X^2 \\ X^{3.4} & X^{3.4} & X^{0.6} & X^{0.6} & X^{3.4} & X^{0.6} & X^{0.6} & X^{3.4} & X^{0.6} & X^{0.6} & X^{3.4} & X^{3.4} & X^{0.6} & X^{3.4} & X^{3.4} & X^{0.6} \\ 1 & X^2 & X^4 & X^2 & X^2 & 1 & X^2 & X^4 & X^4 & X^2 & 1 & X^2 & X^2 & X^4 & X^2 & 1 \\ X^{3.4} & X^{3.4} & X^{0.6} & X^{0.6} & X^{0.6} & X^{3.4} & X^{3.4} & X^{0.6} & X^{0.6} & X^{0.6} & X^{3.4} & X^{3.4} & X^{3.4} & X^{0.6} & X^{0.6} & X^{3.4} \\ X^{0.6} & X^{3.4} & X^{3.4} & X^{0.6} & X^{3.4} & X^{3.4} & X^{0.6} & X^{0.6} & X^{3.4} & X^{0.6} & X^{0.6} & X^{3.4} & X^{0.6} & X^{0.6} & X^{3.4} & X^{3.4} \\ X^2 & 1 & X^2 & X^4 & 1 & X^2 & X^4 & X^2 & X^2 & X^4 & X^2 & 1 & X^4 & X^4 & X^2 & X^2 \\ X^{3.4} & X^{0.6} & X^{0.6} & X^{3.4} & X^{3.4} & X^{3.4} & X^{0.6} & X^{0.6} & X^{0.6} & X^{3.4} & X^{3.4} & X^{0.6} & X^{0.6} & X^{0.6} & X^{3.4} & X^{3.4} \\ 1 & X^2 & X^4 & X^2 & X^2 & 1 & X^2 & X^4 & X^4 & X^2 & 1 & X^2 & X^2 & X^4 & X^2 & 1 \end{array} \right) \end{array} \qquad (5.13)$$

The size of the matrix $\mathbf{B}$ quickly becomes unmanageable, since it grows with the square of the number of states in the code FSM M. As an example, the matrix $\mathbf{B}$ for the four-state 8-PSK trellis code with $h^{(0)} = 5$, $h^{(1)} = 4$, $h^{(2)} = 2$ already has 16×16 entries[2] and is given by (5.13).

We wish to partition our matrix $\mathbf{B}$ into a diverging, a parallel, and a merging section, corresponding to these different stages of an error event. For the four-state code from above this partition is given by the following matrixes:

$$
\mathbf{D} = \frac{1}{4}
\begin{array}{c}
 \\ 00 \\ 11 \\ 22 \\ 33
\end{array}
\begin{pmatrix}
01 & 02 & 03 & 10 & 12 & 13 & 20 & 21 & 23 & 30 & 31 & 33 \\
X^2 & X^4 & X^2 & X^2 & X^2 & X^4 & X^4 & X^2 & X^2 & X^2 & X^4 & X^2 \\
X^2 & X^4 & X^2 & X^2 & X^2 & X^4 & X^4 & X^2 & X^2 & X^2 & X^4 & X^2 \\
X^2 & X^4 & X^2 & X^2 & X^2 & X^4 & X^4 & X^2 & X^2 & X^2 & X^4 & X^2 \\
X^2 & X^4 & X^2 & X^2 & X^2 & X^4 & X^4 & X^2 & X^2 & X^2 & X^4 & X^2
\end{pmatrix},
\tag{5.14}
$$

$$
\mathbf{P} = \frac{1}{4}
\begin{array}{c}
 \\ 01 \\ 02 \\ 03 \\ 10 \\ 12 \\ 13 \\ 20 \\ 21 \\ 23 \\ 30 \\ 31 \\ 32
\end{array}
\begin{pmatrix}
01 & 02 & 03 & 10 & 12 & 13 & 20 & 21 & 23 & 30 & 31 & 33 \\
X^{0.6} & X^{0.6} & X^{3.4} & X^{3.4} & X^{0.6} & X^{0.6} & X^{0.6} & X^{3.4} & X^{0.6} & X^{0.6} & X^{0.6} & X^{3.4} \\
1 & X^2 & X^4 & 1 & X^4 & X^2 & X^2 & X^4 & 1 & X^4 & X^2 & 1 \\
X^{3.4} & X^{3.4} & X^{0.6} & X^{3.4} & X^{0.6} & X^{0.6} & X^{3.4} & X^{0.6} & X^{3.4} & X^{0.6} & X^{0.6} & X^{3.4} \\
X^{3.4} & X^{0.6} & X^{0.6} & X^{0.6} & X^{3.4} & X^{0.6} & X^{0.6} & X^{0.6} & X^{3.4} & X^{3.4} & X^{0.6} & X^{0.6} \\
X^{3.4} & X^{0.6} & X^{0.6} & X^{3.4} & X^{0.6} & X^{3.4} & X^{0.6} & X^{0.6} & X^{3.4} & X^{0.6} & X^{3.4} & X^{3.4} \\
1 & X^2 & X^4 & 1 & X^4 & X^2 & X^2 & X^2 & 1 & X^4 & X^2 & 1 \\
1 & X^2 & X^4 & 1 & X^4 & X^2 & X^2 & X^4 & 1 & X^4 & X^2 & 1 \\
X^{3.4} & X^{0.6} & X^{0.6} & X^{3.4} & X^{0.6} & X^{3.4} & X^{0.6} & X^{0.6} & X^{3.4} & X^{0.6} & X^{3.4} & X^{3.4} \\
X^{3.4} & X^{0.6} & X^{0.6} & X^{0.6} & X^{3.4} & X^{0.6} & X^{0.6} & X^{0.6} & X^{3.4} & X^{3.4} & X^{0.6} & X^{0.6} \\
X^{3.4} & X^{3.4} & X^{0.6} & X^{3.4} & X^{0.6} & X^{0.6} & X^{3.4} & X^{0.6} & X^{3.4} & X^{0.6} & X^{0.6} & X^{3.4} \\
1 & X^2 & X^4 & 1 & X^4 & X^2 & X^2 & X^4 & 1 & X^4 & X^4 & X^2 \\
X^{0.6} & X^{0.6} & X^{3.4} & X^{3.4} & X^{0.6} & X^{0.6} & X^{0.6} & X^{3.4} & X^{0.6} & X^{0.6} & X^{0.6} & X^{3.4}
\end{pmatrix},
\tag{5.15}
$$

---

[2] The exponents have been rounded to one decimal place for space reasons.

$$
\mathbf{M} = \frac{1}{4}
\begin{array}{c}
\\ 01 \\ 02 \\ 03 \\ 10 \\ 12 \\ 13 \\ 20 \\ 21 \\ 23 \\ 30 \\ 31 \\ 32
\end{array}
\begin{pmatrix}
00 & 11 & 22 & 33 \\
X^{3.4} & X^{3.4} & X^{3.4} & X^{3.4} \\
X^{2} & X^{2} & X^{2} & X^{2} \\
X^{0.6} & X^{3.4} & X^{0.6} & X^{3.4} \\
X^{3.4} & X^{3.4} & X^{3.4} & X^{3.4} \\
X^{3.4} & X^{0.6} & X^{3.4} & X^{0.6} \\
X^{2} & X^{2} & X^{2} & X^{2} \\
X^{2} & X^{2} & X^{2} & X^{2} \\
X^{3.4} & X^{0.6} & X^{3.4} & X^{0.6} \\
X^{3.4} & X^{3.4} & X^{3.4} & X^{3.4} \\
X^{0.6} & X^{3.4} & X^{0.6} & X^{3.4} \\
X^{2} & X^{2} & X^{2} & X^{2} \\
X^{3.4} & X^{3.4} & X^{3.4} & X^{3.4}
\end{pmatrix},
\tag{5.16}
$$

where the diverging matrix $\mathbf{D}$ is $N \times (N^2 - N)$, the parallel matrix $\mathbf{P}$ is $(N^2 - N) \times (N^2 - N)$, and the merging matrix $\mathbf{M}$ is $(N^2 - N) \times N$.

From Figure 5.5 we see that each error event starts when the correct path and the error path diverge and terminates when the two paths merge again. In between e_i and c are never in the same state at the same time unit, and we refer to this middle part of the error event as the parallel section.

We may now describe all distances of error events of length exactly L by

$$
\mathbf{G}_L = \mathbf{D}\mathbf{P}^{L-2}\mathbf{M}, \qquad L \geq 2.
\tag{5.17}
$$

The $((p, p), (q, q))$-entry of $\mathbf{G}_L$ is essentially a table of all weighted distances between length-L path pairs that originate from a given state p and merge in a given state q, L time units later. Equation (5.17) can now be used to find the distance spectrum of a code up to any desired length. Although (5.17) is a convenient way to describe the distance spectrum, its calculation is best performed using an adaptation of one of the decoding algorithms described in Chapter 6.

5.3   THE TRANSFER FUNCTION BOUND

The union bounds (5.9) and (5.10) are very tight for error probabilities smaller than about 10^{-2}, as evidenced in Figure 5.4. Its disadvantage is that we need to evaluate the distance spectrum, and we also need some way of terminating the sum since (5.9) and (5.10) have no closed-form expressions.

Not so for our next bound, which is traditionally referred to as the *transfer function bound* since it can also be derived via state transfer functions from a systems theory point of view. The transfer function bound using the pair-state trellis was first applied to TCM in [2]. The first step we take is loosening (5.9) by applying the inequality (see, e.g., [3, page 83])

$$Q\left(\sqrt{d_i^2 \frac{RE_b}{2N_0}}\right) \le \exp\left(-d_i^2 \frac{RE_b}{4N_0}\right). \tag{5.18}$$

Now, using (5.17), which is essentially a table of the distances between path pairs of length L, it is easy to see that

$$\overline{P_e} \le \sum_{\substack{i \\ (d_i^2 \in \mathcal{D})}} A_{d_i^2} \exp\left(-\sqrt{d_i^2 \frac{RE_b}{2N_0}}\right) \tag{5.19}$$

$$= \frac{1}{N} \sum_{L=2}^{\infty} \mathbf{1}^T \mathbf{D} \mathbf{P}^{L-2} \mathbf{M} \ \mathbf{1} \Bigg|_{X=\exp(-RE_b/4N_0)}, \tag{5.20}$$

where $\mathbf{1} = (1, 1, \ldots, 1)^T$ is an $N \times 1$ vector of all ones and acts by summing all merger states while $\frac{1}{N}\mathbf{1}^T$ averages uniformly over all diverging states.

Pulling the sum into the matrix multiplication, we obtain

$$\overline{P_e} \le \frac{1}{N} \mathbf{1}^T \mathbf{D} \sum_{L=2}^{\infty} \mathbf{P}^{L-2} \mathbf{M} \ \mathbf{1} \Bigg|_{X=\exp(-RE_b/4N_0)} \tag{5.21}$$

$$= \frac{1}{N} \mathbf{1}^T \mathbf{D} (\mathbf{I} - \mathbf{P})^{-1} \mathbf{M} \ \mathbf{1} \Bigg|_{X=\exp(-RE_b/4N_0)}. \tag{5.22}$$

Equation (5.22) involves the inversion[3] of an $(N^2 - N) \times (N^2 - N)$ matrix, which might be a sizable task. In going from (5.21) to (5.22) we assumed that the infinite series in (5.21) converges. With matrix theory it can be shown that this series converges if the largest eigenvalue of $\mathbf{P}$, $\lambda_{\max} < 1$ [5], and hence $(\mathbf{I} - \mathbf{P})^{-1}$, exists. This is always the case for noncatastrophic codes and for E_b/N_0 sufficiently large.

Note also that it is necessary to invert this matrix symbolically in order to obtain a closed-form expression for the transfer function bound as a function of the signal-to-noise ratio E_b/N_0.

---

[3] The matrix $\mathbf{I} - \mathbf{P}$ is sparse in most cases of interest and can therefore be inverted efficiently numerically [4].

Since using (5.17) to search for all the distances up to a certain length L_{max} is often easier, it is more efficient to use the tighter union bound for the distances up to a certain path pair length L_{max} and use the transfer function only to bound the tail; that is, we break (5.17) into two components and obtain

$$\overline{P_e} \leq \sum_{L=1}^{L_{max}} \sum_{c_k^{(L)}} p(c_k^{(L)}) \sum_{e_i^{(L)}|c_k^{(L)}} Q\left(\sqrt{d_{ci}^2 \frac{RE_b}{2N_0}}\right)$$

$$+ \frac{1}{N}\mathbf{1}^T\mathbf{D} \sum_{L=L_{max}+1}^{\infty} \mathbf{P}^{L-2}\mathbf{M} \ \ \mathbf{1}\Bigg|_{X=\exp(-RE_b/4N_0)}, \tag{5.23}$$

where $c^{(L)}$ and $e_i^{(L)}$ are a correct path and incorrect path of length exactly L. The second term, the tail of the transfer function bound, can be overbounded by

$$\epsilon = \frac{1}{N}\mathbf{1}^T\mathbf{D} \sum_{L=L_{max}+1}^{\infty} \mathbf{P}^{L-2}\mathbf{M} \ \ \mathbf{1}\Bigg|_{X=\exp(-RE_b/4N_0)} \leq \frac{\lambda_{max}^{L_{max}-1}}{1-\lambda_{max}}, \tag{5.24}$$

where λ_{max} is the largest eigenvalue[4] of $\mathbf{P}$.

The computation of the eigenvalues of a matrix and its inversion are closely related, and not much seems to have been gained. The complexities of inverting $\mathbf{I} - \mathbf{P}$ and finding λ_{max} are comparable. However, there are many efficient ways of bounding λ_{max}, and we do not even need a very tight bound since we are interested only in the order of magnitude of ϵ.

One straightforward way of bounding λ_{max} that we will use is obtained by applying Gerschgorin's circle theorem (see Appendix 5.A). Doing this we find that

$$\lambda_{max} \leq \max_i \sum_j p_{ij} = g; \tag{5.25}$$

i.e., λ_{max} is smaller than the the maximum row sum of $\mathbf{P}$. A quick look at (5.15) shows that g, which in general is a polynomial in X, contains the constant term 1 for our four-state code. This means that $g \geq 1$ for all real X and (5.25) is not useful as a bound in (5.24).

Since working with $\mathbf{P}$ did not lead to a satisfactory bound, we go to $\mathbf{P}^2$. From basic linear algebra we know that the maximum eigenvalue of $\mathbf{P}^2$ is

---

[4] The Perron-Frobenious theorem from linear algebra tells us that a nonnegative matrix has a nonnegative largest eigenvalue [5]. A matrix is called nonnegative if every entry is a nonnegative real number, which is the case for $\mathbf{P}$.

λ^2_{max}. Let us then apply the Gerschgorin bound to $\mathbf{P}^2$ and obtain

$$\lambda_{max} \leq \sqrt{\max_i \sum_j p^{(2)}_{ij}} = \sqrt{g^{(2)}}. \tag{5.26}$$

Figure 5.6 shows plots of

$$\left(\sqrt{g^{(2)}}\right)^{L_{max}-1} \Big/ \left(1 - \sqrt{g^{(2)}}\right) \tag{5.27}$$

versus the signal-to-noise ratio E_b/N_0 for various maximum lengths L_{max}.

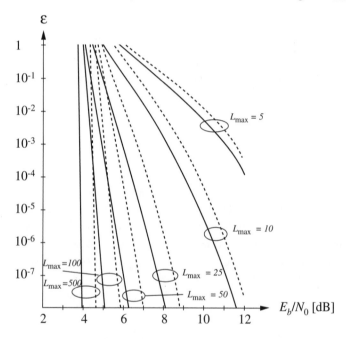

Figure 5.6 Plots of the bound on ϵ (5.24) using the exact value of λ_{max} (solid curves) as well as the Gerschgorin bound (dotted curves) for the four-state 8-PSK trellis code for $L_{max} = 5, 10, 25, 50, 100,$ and 500.

From such plots one can quickly determine the maximum length L_{max} in the trellis that needs to be searched for the distances d^2_{ci} in order to achieve a prescribed accuracy.

5.4   REDUCTION THEOREMS

The reader may well wonder at this stage about the complexity of our error calculation, since N^2 grows quickly beyond convenience and makes the

Since using (5.17) to search for all the distances up to a certain length L_{max} is often easier, it is more efficient to use the tighter union bound for the distances up to a certain path pair length L_{max} and use the transfer function only to bound the tail; that is, we break (5.17) into two components and obtain

$$\overline{P_e} \leq \sum_{L=1}^{L_{max}} \sum_{c_k^{(L)}} p(c_k^{(L)}) \sum_{e_i^{(L)} | c_k^{(L)}} Q\left(\sqrt{d_{ci}^2 \frac{RE_b}{2N_0}}\right)$$
$$+ \frac{1}{N} \mathbf{1}^T \mathbf{D} \sum_{L=L_{max}+1}^{\infty} \mathbf{P}^{L-2} \mathbf{M} \left.\mathbf{1}\right|_{X=\exp(-RE_b/4N_0)}, \tag{5.23}$$

where $c^{(L)}$ and $e_i^{(L)}$ are a correct path and incorrect path of length exactly L. The second term, the tail of the transfer function bound, can be overbounded by

$$\epsilon = \frac{1}{N} \mathbf{1}^T \mathbf{D} \sum_{L=L_{max}+1}^{\infty} \mathbf{P}^{L-2} \mathbf{M} \left.\mathbf{1}\right|_{X=\exp(-RE_b/4N_0)} \leq \frac{\lambda_{max}^{L_{max}-1}}{1 - \lambda_{max}}, \tag{5.24}$$

where λ_{max} is the largest eigenvalue[4] of $\mathbf{P}$.

The computation of the eigenvalues of a matrix and its inversion are closely related, and not much seems to have been gained. The complexities of inverting $\mathbf{I} - \mathbf{P}$ and finding λ_{max} are comparable. However, there are many efficient ways of bounding λ_{max}, and we do not even need a very tight bound since we are interested only in the order of magnitude of ϵ.

One straightforward way of bounding λ_{max} that we will use is obtained by applying Gerschgorin's circle theorem (see Appendix 5.A). Doing this we find that

$$\lambda_{max} \leq \max_i \sum_j p_{ij} = g; \tag{5.25}$$

i.e., λ_{max} is smaller than the the maximum row sum of $\mathbf{P}$. A quick look at (5.15) shows that g, which in general is a polynomial in X, contains the constant term 1 for our four-state code. This means that $g \geq 1$ for all real X and (5.25) is not useful as a bound in (5.24).

Since working with $\mathbf{P}$ did not lead to a satisfactory bound, we go to $\mathbf{P}^2$. From basic linear algebra we know that the maximum eigenvalue of $\mathbf{P}^2$ is

---

[4] The Perron-Frobenius theorem from linear algebra tells us that a nonnegative matrix has a nonnegative largest eigenvalue [5]. A matrix is called nonnegative if every entry is a nonnegative real number, which is the case for $\mathbf{P}$.

$\lambda_{\max}^2$. Let us then apply the Gerschgorin bound to $\mathbf{P}^2$ and obtain

$$\lambda_{\max} \le \sqrt{\max_i \sum_j p_{ij}^{(2)}} = \sqrt{g^{(2)}}. \tag{5.26}$$

Figure 5.6 shows plots of

$$\left(\sqrt{g^{(2)}}\right)^{L_{\max}-1} \bigg/ \left(1 - \sqrt{g^{(2)}}\right) \tag{5.27}$$

versus the signal-to-noise ratio E_b/N_0 for various maximum lengths $L_{\max}$.

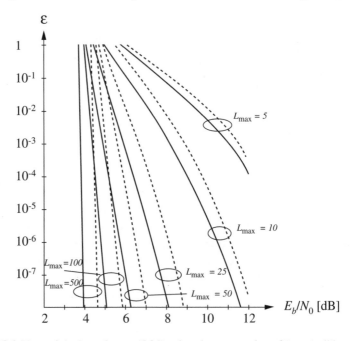

Figure 5.6 Plots of the bound on ϵ (5.24) using the exact value of $\lambda_{\max}$ (solid curves) as well as the Gerschgorin bound (dotted curves) for the four-state 8-PSK trellis code for $L_{\max} = 5, 10, 25, 50, 100,$ and 500.

From such plots one can quickly determine the maximum length $L_{\max}$ in the trellis that needs to be searched for the distances d_{ci}^2 in order to achieve a prescribed accuracy.

5.4   REDUCTION THEOREMS

The reader may well wonder at this stage about the complexity of our error calculation, since N^2 grows quickly beyond convenience and makes the evaluation of (5.9) or (5.22) a task for supercomputers. Fortunately there are large classes of trellis codes for which this error computation can be simplified considerably.

For convolutional codes, for instance, we know that linearity (see Definition 4.1 and preceding discussion) implies that the choice of the correct sequence c does not matter. The same is true for the geometrically uniform codes discussed in Section 3.8. In fact, for any regular code, the distance spectrum is the same for every correct path (codeword), and we may therefore assume c to be any convenient path (codeword), say the all-zero path, and do this without loss of generality. This corresponds to a reduction in the size of $\mathbf{P}$ to $N - 1 \times N - 1$, since averaging over c in (5.8) is not required.

For general trellis codes this simplification is not so obvious. Thus, to proceed we need the notion of equivalence of finite-state machines and their states. The idea behind this approach is to replace our $(N^2 - N)$-state FSM $\mathcal{M}$ by a possibly much smaller FSM, one with fewer states.

The following two definitions will be used later:

DEFINITION 5.1

Two finite-state machines $\mathcal{M}_1$ and $\mathcal{M}_2$ are called *output equivalent* if and only if their output sequences are identical for identical input sequences and equivalent starting states.

DEFINITION 5.2

Two states u and v of a FSM $\mathcal{M}$ are output equivalent, denoted by $u \equiv v$, if and only if

i   for every transition $u \to u'$, there exists a transition $v \to v'$ such that the outputs $\delta(u \to u') = \delta(v \to v')$, and vice versa, and

ii the corresponding successor states are also equivalent; that is, $u' \equiv v'$.

Definition 5.2 guarantees that if $u \equiv v$, the sets of possible output sequences from u and v are identical. We may therefore combine u and v into a single state w that accounts for both u and v and thus generate an equivalent FSM with one state less than $\mathcal{M}$. For the general case there exists a partition algorithm that converges in a finite number of steps to an equivalent FSM having the minimum number of states. This algorithm is not needed for our further discussion here and is described in Appendix 5.B.

We are now ready for our first reduction, using the quasi-regularity of certain signal sets, as discussed in Definition 3.2.

THEOREM 5.1

The number of states in the FSM $\mathcal{M}$ generating the distances d_i^2 and their multiplicities $A_{d_i^2}$ in group-trellis codes that use quasi-regular signal sets can be reduced from N^2 to N.

PROOF

The output signal of a TCM code is given by the mapper function $\tau(u, s_i)$. Since we assume group-trellis codes (Section 3.2), the set of states s_i forms an additive group and we may pick $p = s_i$ and $q = s_i \oplus e_s$. The output δ of $\mathcal{M}$ from pair-state $(p, q) = (s_i, s_i \oplus e_s)$ with inputs u and $u \oplus e_u$ is the Euclidean distance $d^2 = \|\tau(u, s_i) - \tau(u \oplus e_u, s_i \oplus e_s)\|^2$, as shown in Figure 5.7.

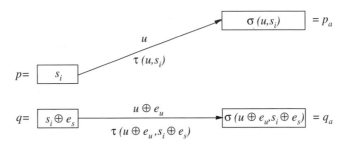

Figure 5.7 Inputs and outputs for the states s_i and $s_i \oplus e_s$ of an Ungerböck trellis code.

The set of distances $\{d_i^2\}$ from the state $(s_i, s_i \oplus e_s)$ as we vary over all u is $\{d_{e_u, e_s}^2\}$ and is independent of s_i for quasi-regular signal sets from Definition 3.2. For quasi-regular signal sets, the same set of distances originates from state $(p', q') = (s_k, s_k \oplus e_s)$ as from (p, q), where s_k is arbitrary, establishing (i) in Definition 5.2.

Since $\{d_{e_u, e_s}^2\}$ is independent of s_i there exists for each u a unique u' such that

$$\|\tau(u, s_i) - \tau(u \oplus e_u, s_i \oplus e_s)\|^2 = \|\tau(u', s_k) - \tau(u' \oplus e_u, s_k \oplus e_s)\|^2$$
$$(5.28)$$
$$= \delta_a.$$

The successors of the associated transitions are $(p_a, q_a) = (\sigma(u, s_i), \sigma(u \oplus e_u, s_i \oplus e_s))$ and $(p'_a, q'_a) = (\sigma(u', s_k), \sigma(u' \oplus e_u, s_k \oplus e_s))$ (see Figure 5.8). $\sigma(u, s_i)$ denotes the state-transition function that gives a successor state as a function of the previous state s_i and the input u.

But due to the additivity of the trellis $(p_a, q_a) = (\sigma(u, s_i), \sigma(u \oplus e_u, s_i \oplus e_s)) = (s_{i'}, s_{i'} \oplus e_{s'})$ implies $(p'_a, q'_a) = (\sigma(s_k, u'), \sigma(s_k \oplus e_s, u' \oplus e_u)) = (s_{k'}, s_{k'} \oplus e_{s'})$. Therefore $(s_i, s_i \oplus e_s) \equiv (s_k, s_k \oplus e_s)$ since their successors

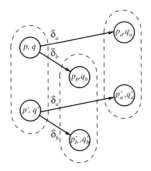

Figure 5.8 Illustration of state equivalence.

$(p_a, q_a) \equiv (p'_a, q'_a)$ for every output distance δ_a by induction on u and the length of the trellis.

Since there are $N - 1$ states $s_k \neq s_i$ there are N equivalent states $(s_k, s_k \oplus e_s)$, $s_k \in M$. Each class of equivalent states can be represented by e_s with outputs $\{d_{e_s}, e_u\}$ and successors $(\sigma(e_u, e_s))$.

COROLLARY 5.2

The number of states in the FSM $\mathcal{M}$ for the MPSK and QAM codes in Section 3.4 (Ungerböck TCM codes) can be reduced from N^2 to N.

Proof

Ungerböck codes are group-trellis codes from Section 3.2, using quasi-regular signal constellations. We have already shown in Theorem 3.1 that MPSK signal sets are quasi-regular. Quasi-regularity can also be shown for the QAM signal sets used in the Ungerböck codes. This is a tedious, albeit straightforward, exercise.

As an example, consider again the four-state trellis code from page 128. The states $00 \equiv 11 \equiv 22 \equiv 33$ are all equivalent and their equivalent state is $e_s = 0$. Similarly, $01 \equiv 10 \equiv 23 \equiv 32$, $02 \equiv 13 \equiv 20 \equiv 31$, and $03 \equiv 12 \equiv 21 \equiv 30$ with equivalent states $e_s = 1$, 2, and 3, respectively.

The reduced FSM can now be obtained via the following simple procedure. Any equivalent state can be used to describe the distances for the entire equivalence class. We therefore eliminate all but one equivalent state in each class, and we choose this state to be e_s. This amounts to canceling the corresponding rows in **B**. Next, mergers into a canceled state have to be handled by its equivalent state. We therefore add all the columns of canceled equivalent states to the one column of e_s, producing as entries polynomials in X with exponents from the set $\{d_{e_s, e_u}\}$.

Doing this with the 16×16 matrix (5.13), we obtain the reduced transfer matrix

$$\mathbf{B}_r = \begin{pmatrix} 1 & X^2 & X^4 & X^2 \\ X^{3.41} & \frac{1}{2}X^{0.59}+\frac{1}{2}X^{3.41} & X^{0.59} & \frac{1}{2}X^{0.59}+\frac{1}{2}X^{3.41} \\ X^2 & 1 & X^2 & X^4 \\ \frac{1}{2}X^{0.59}+\frac{1}{2}X^{3.41} & X^{3.41} & \frac{1}{2}X^{0.59}+\frac{1}{2}X^{3.41} & X^{0.59} \end{pmatrix}, \quad (5.29)$$

which we can likewise partition into a reduced diverging matrix $\mathbf{D}_r$, a reduced parallel matrix $\mathbf{P}_r$, and a reduced merging matrix $\mathbf{M}_r$. We obtain

$$\mathbf{D}_r = \begin{pmatrix} X^2 & X^4 & X^2 \end{pmatrix}, \quad (5.30)$$

$$\mathbf{P}_r = \begin{pmatrix} X^2 & X^4 & X^2 \\ \frac{1}{2}X^{0.59}+\frac{1}{2}X^{3.41} & X^{0.59} & \frac{1}{2}X^{0.59}+\frac{1}{2}X^{3.41} \\ 1 & X^2 & X^4 \\ X^{3.41} & \frac{1}{2}X^{0.59}+\frac{1}{2}X^{3.41} & X^{0.59} \end{pmatrix}, \quad (5.31)$$

$$\mathbf{M}_r = \begin{pmatrix} X^{3.41} \\ X^{0.59} \\ \frac{1}{2}X^{0.59}+\frac{1}{2}X^{3.41} \end{pmatrix}. \quad (5.32)$$

The transfer function bound (5.20) can now be simplified to

$$\overline{P}_e \leq \sum_{L=2}^{\infty} \mathbf{D}_r \mathbf{P}_r^{L-2} \mathbf{M}_r \Bigg|_{X=\exp\left(-d_i^2 RE_b/4N_0\right)} ; \quad (5.33)$$

i.e., the summation over all starting and terminating states could be eliminated. Finding the distance spectrum of Ungerböck codes becomes essentially as simple as finding the distance spectrum of regular or geometrically uniform codes. Remember that convolutional codes, which are linear in the output space, are the best-known representatives of regular codes.

Theorem 5.1 is valid for the distance spectrum as used in (5.9). The following corollary ensures us that the bit error multiplicities can also be calculated by a reduced FSM.

COROLLARY 5.3

The number of states in the FSM $\mathcal{M}$ describing the distances d_i^2 and bit error multiplicities $B_{d_i^2}$ in Ungerböck TCM codes can be reduced from N^2 to N.

PROOF

In order to keep track of the bit errors, we have to add the bit error weight to the output of $\mathcal{M}$, i.e., the output is given by the pair $\delta((p, q) \to (p_1, q_1)) = \left(d_{(p,q),(p_1,q_1)}^2, w(e_u)\right)$, where $w(e_u)$ is the Hamming weight of e_u and e_u is the bit error pattern between the transitions $p \to p_1$ and $q \to q_1$. But extending

(5.28) leads to

$$\left(\| \tau(s_i, u) - \tau(s_i \oplus e_s, u \oplus e_u) \|^2, w(e_u) \right)$$

$$= \left(\| \tau(s_k, u') - \tau(s_k \oplus e_s, u' \oplus e_u) \|^2, w(e_u) \right) \qquad (5.34)$$

$$= (\delta_a, w(e_u)),$$

which is also independent of s_i as we vary over all u. The rest of the proof is exactly as in Theorem 5.1.

The approach chosen in this section does not reflect the actual historical development of these results. Methods to overcome the nonlinearity of Ungerböck trellis codes were first presented by Zehavi and Wolf [6] and, independently, Rouanne and Costello [7] with generalizations in [8] and appeared under the suggestive term "quasi-regularity." Later a slightly different, but equivalent, approach was given by Biglieri and McLane [9], who called the property necessary for reduction "uniformity."

We have chosen the state-reduction point of view both because it appeared to us to be easier to understand as well as more general. Reduction theorems have also been proven for continuous-phase frequency-shift keying [1] and complementary convolutional codes [10], where the number of equivalent states could be reduced to less than the number of states in the code FSM M.

5.5 RANDOM CODING BOUNDS

In this section we study random coding bounds for trellis codes. The idea of random coding originated from Shannon and is, loosely speaking, this: The calculation of the error probability is very difficult for any code of decent size. Furthermore, there is little hope of calculating an exact error probability for these codes, and finding bounds on their performance becomes the next best thing one can hope for. But, as we have seen, even finding bounds for specific codes is rather cumbersome.

We need to know the distance spectrum of the code as discussed earlier. If we average the performance (i.e., the block) bit, or event error probabilities over all possible codes, assuming some distribution on the selection of codes, we find to our great astonishment that this average can be calculated relatively easily.

Knowing the average of the performance of all codes, we can be rest assured that there must exist at least one code whose individual performance is better than that calculated average. This leads to the famous existence argument of good codes and immediately to the dilemma that, while knowing of the existence of good codes, the argument tells us nothing about how to find one.

Random coding bounds are still of interest, even if they do not give us a method for finding good codes. The theorems tell us about the performance of good codes and give us a way to evaluate signal constellations. In this section we talk about random coding bounds on the first event error probability of trellis codes.

Specifically, let us assume that we use a maximum-likelihood (ML) decoder that selects the hypothesized transmitted sequence $\hat{\mathbf{x}}$ according to the highest probability metric $p(\mathbf{y}|\hat{\mathbf{x}})$, where $\mathbf{y}$ is the received sequence $\mathbf{x} + \mathbf{n}$, $\mathbf{x}$ is the transmitted sequence, and $\mathbf{n}$ is a noise sequence (see Section 2.5). This decoder makes an error if it decodes a sequence $\mathbf{x}'$, given that the transmitted sequence was $\mathbf{x}$. This happens if $p(\mathbf{y}|\mathbf{x}) \le p(\mathbf{y}|\mathbf{x}')$.

Let c and e be paths through the trellis. Remember that c and e describe the paths through the trellis (state sequences) but not the signals on these paths. The signal assignment is the actual encoding function, and we average over all these functions. As with specific trellis codes, let c be the correct path (the one taken by the encoder), and let e denote the incorrect path that diverges from c at node j. Further, let $\mathcal{E}$ denote the set of all incorrect paths that diverge from c at node j.

If we are using a group-trellis code, the sets $\mathcal{E}$ for different correct paths are identical. They contain the same number of paths of a certain length. In a particular trellis code, then, $\mathbf{x}$ is the sequence of signals assigned along the correct path c and $\mathbf{x}'$ is the sequence of signals assigned to e.

A necessary condition for an error event to start at node j is that the incorrect path e accumulate a higher total probability than the correct path c over their unmerged segments of the trellis.

Using this notation we rewrite (5.5) as

$$\overline{P_e} \le \sum_c p(c) \int_{\mathbf{v}} p(\mathbf{y}|\mathbf{x}) \mathcal{I} \left(\bigcup_{e \in \mathcal{E}} e(p(\mathbf{y}|\mathbf{x}') \ge p(\mathbf{y}|\mathbf{x})) \right) d\mathbf{y}, \qquad (5.35)$$

where $\mathcal{I}(B)$ is an indicator function such that $\mathcal{I}(B) = 0$ if $B = \emptyset$ (the empty set) $\mathcal{I}(B) = 1$ if $B \ne \emptyset$, and $e(p(\mathbf{y}|\mathbf{x}') \ge p(\mathbf{y}|\mathbf{x}))$ is a path e for which $p(\mathbf{y}|\mathbf{x}') \ge p(\mathbf{y}|\mathbf{x})$. This then also implies that we are using an ML decoder (see Section 2.4). $\mathcal{I}\left(\bigcup_{e \in \mathcal{E}} e(p(\mathbf{y}|\mathbf{x}') \ge p(\mathbf{y}|\mathbf{x}))\right)$ simply says whether or not there exists an error path with a metric smaller than the correct path.

We now specify the structure of our group-trellis code to be the one shown in Figure 5.9. The binary input vector at time r, u_r, has k components

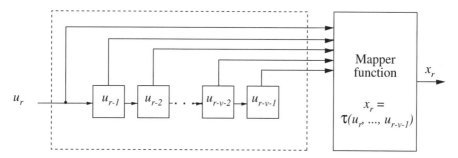

Figure 5.9 Group-trellis encoder structure used for the random coding bound
arguments.

and enters a feedforward shift register at time unit r. The shift register stores
the $\nu - 1$ most recent binary input vectors. The mapper function assigns a
symbol x_r as a function of the input and the state; that is, $x_r = \tau(u_r, s_r) = \tau(u_r, \ldots, u_{r-\nu-1})$.

The symbol x_r is chosen from a signal set $\mathcal{A}$ of size A. The channel
considered here is a memoryless channel used without feedback, that is,
the output symbol y is described by the conditional probability distribution
$p(y|x)$, and the conditional probability of a symbol vector $p(\mathbf{y}|\mathbf{x})$ is then
the product of the individual conditional symbol probabilities. We will
need this property in the proof to obtain an exponential bound on the error
probability.

The particular realization of the group-trellis code as in Figure 5.9
has been chosen merely for convenience. The bounds that follow can be
calculated for other structures as well, but the formalism can become quite
unpleasant for some of them. Note that this structure is a group-trellis
encoder as introduced in Figure 3.6.

For an incorrect path e, which diverges from c at node j, to merge with
the correct path at node $j + L$, the last $\nu - 1$ entries in the information
sequence $u'_j, \ldots, u'_{j+L}$ associated with e must equal the last $\nu - 1$ entries in
the information sequences $u_j, \ldots, u_{j+L}$ associated with c. This must be so
because paths e and c merge at node $j + L$ only if their associated encoder
states are identical. Because an information vector u_j entering the encoder
can affect the output for ν time units, this is also the time it takes to force
the encoder into any given state from any arbitrary starting state. Because
the remaining information bits are arbitrary, we have $N_p \leq (2^k - 1)2^{k(L-\nu)}$
incorrect paths e of length L. Note that the choice of the information bits
at node j is restricted because we stipulated that the incorrect path diverges
at node j, which rules out the one path that continues to the correct state at
node $j + 1$. This accounts for the factor $2^k - 1$ in the expression for N_p.

We now proceed to express (5.35) as the sum over sequences of length L, and rewrite (5.35) as

$$\overline{P_e} \leq \sum_{L=v}^{\infty} \sum_{c^{(L)} \in \mathcal{C}^{(L)}} p(c^{(L)}) \int_{\mathbf{y}} p(\mathbf{y}|\mathbf{x}) \mathcal{I} \left(\bigcup_{e^{(L)} \in \mathcal{E}^{(L)}} e^{(L)}(p(\mathbf{y}|\mathbf{x}) \leq p(\mathbf{y}|\mathbf{x}')) \right) d\mathbf{y}, \quad (5.36)$$

where $\mathcal{C}^{(L)}$ is the set of all correct paths $c^{(L)}$ of length L starting at node j and $\mathcal{E}^{(L)}$ is the set of all incorrect paths $e^{(L)}$ of length L unmerged with $c^{(L)}$ from node j to node $j + L$. Note that $\bigcup_L \mathcal{E}^{(L)} = \mathcal{E}$.

Now we observe that if an error occurs at node j, then for at least one path e

$$p(\mathbf{y}|\mathbf{x}') \geq p(\mathbf{y}|\mathbf{x}). \quad (5.37)$$

Therefore, for any real parameter $\gamma \geq 0$,

$$\sum_{e^{(L)} \in \mathcal{E}^{(L)}} \left[\frac{p(\mathbf{y}|\mathbf{x}')}{p(\mathbf{y}|\mathbf{x})} \right]^{\gamma} \geq 1. \quad (5.38)$$

We can raise both sides of (5.38) to the power of some nonnegative parameter $\rho \geq 0$ and preserve the inequality:

$$\left(\sum_{e^{(L)} \in \mathcal{E}^{(L)}} \left[\frac{p(\mathbf{y}|\mathbf{x}')}{p(\mathbf{y}|\mathbf{x})} \right]^{\gamma} \right)^{\rho} \geq 1. \quad (5.39)$$

We now use (5.39) to bound the indicator function $\mathcal{I}(\cdot)$ above by an exponential:

$$\mathcal{I} \left(\bigcup_{e^{(L)} \in \mathcal{E}^{(L)}} e^{(L)}(p(\mathbf{y}|\mathbf{x}) \leq p(\mathbf{y}|\mathbf{x}')) \right) \leq \left(\sum_{e^{(L)} \in \mathcal{E}^{(L)}} \left[\frac{p(\mathbf{y}|\mathbf{x}')}{p(\mathbf{y}|\mathbf{x})} \right]^{\gamma} \right)^{\rho}. \quad (5.40)$$

Using (5.40) in (5.36) we obtain

$$\overline{P_e} \leq \sum_{L=v}^{\infty} \sum_{c^{(L)}} p(c^{(L)}) \int_{\mathbf{y}} p(\mathbf{y}|\mathbf{x}) \left(\sum_{e^{(L)} \in \mathcal{E}^{(L)}} \left[\frac{p(\mathbf{y}|\mathbf{x}')}{p(\mathbf{y}|\mathbf{x})} \right]^{\gamma} \right)^{\rho} d\mathbf{y}. \quad (5.41)$$

Let us pause and reflect on what we are doing. If we set the parameter $\rho = 1$, the sum over $e^{(L)}$ will pull out all the way and we have in effect the union bound as in (5.6). We know, however, that the union bound is rather loose for low signal-to-noise ratios or for high rates—hence, the parameter ρ, which allows us to tighten the bound in these areas.

Since we are free to choose γ, we select $\gamma = 1/(1 + \rho)$, which simplifies (5.41) to

$$\overline{P_e} \leq \sum_{L=v}^{\infty} \sum_{c^{(L)}} p(c^{(L)}) \int_{\mathbf{y}} p(\mathbf{y}|\mathbf{x})^{1/(1+\rho)} \left(\sum_{e^{(L)} \in \mathcal{E}^{(L)}} p(\mathbf{y}|\mathbf{x}')^{1/(1+\rho)} \right)^{\rho} d\mathbf{y}. \quad (5.42)$$

$\overline{P_e}$ is the event error probability of a particular code since it depends on the signal sequences $\mathbf{x}$ and $\mathbf{x}'$ of the code. The aim of this section is to obtain a bound on an ensemble average of trellis codes, and we therefore average $\overline{P_e}$ over all the codes in the ensemble:

$$
\text{Avg}\left\{\overline{P_e}\right\}
$$

$$
\leq \text{Avg}\left\{\sum_{L=\nu}^{\infty}\sum_{c^{(L)}} p(c^{(L)})\int_{\mathbf{y}} p(\mathbf{y}|\mathbf{x})^{1/(1+\rho)}\left(\sum_{e^{(L)}\in\mathcal{E}^{(L)}} p(\mathbf{y}|\mathbf{x}')^{1/(1+\rho)}\right)^{\rho} d\mathbf{y}\right\}, \tag{5.43}
$$

where Avg$\{\cdot\}$ denotes this ensemble average.

Using the linearity of the averaging operator and noting that there are exactly $N = 2^{kL}$ equiprobable paths in $\mathcal{C}^{(L)}$ of length L, because at each time unit there are 2^k possible choices to continue the correct path, we obtain

$$
\text{Avg}\left\{\overline{P_e}\right\}
$$

$$
\leq \sum_{L=\nu}^{\infty}\frac{1}{2^{kL}}\text{Avg}\left\{\sum_{c^{(L)}}\int_{\mathbf{y}} p(\mathbf{y}|\mathbf{x})^{1/(1+\rho)}\left(\sum_{e^{(L)}\in\mathcal{E}^{(L)}} p(\mathbf{y}|\mathbf{x}')^{1/(1+\rho)}\right)^{\rho} d\mathbf{y}\right\}. \tag{5.44}
$$

To continue, let $\mathcal{X} = \{\mathbf{x}_1, \mathbf{x}_2, \ldots, \mathbf{x}_N\}$ be a possible assignment of signal sequences of length L associated with the paths $c^{(L)}$ (i.e., $\mathcal{X}$ is a particular code) and let $q_{LN}(\{\mathcal{X}\}) = q_{LN}(\mathbf{x}_1, \ldots, \mathbf{x}_N)$ be the probability of choosing this code. Note that $\mathcal{E}^{(L)} \subset \mathcal{C}^{(L)}$, since each incorrect path is also a possible correct path. Averaging over all codes means averaging over all signal sequences in these codes (i.e., over all assignments $\mathcal{X}$). Doing this we obtain

$$
\text{Avg}\left\{\overline{P_e}\right\} \leq \sum_{L=\nu}^{\infty}\frac{1}{2^{kL}}\sum_{c^{(L)}}\int_{\mathbf{y}}\sum_{\mathbf{x}_1}\cdots\sum_{\mathbf{x}_N} q_{LN}(\mathcal{X})p(\mathbf{y}|\mathbf{x})^{1/(1+\rho)}
$$

$$
\times\left(\sum_{e^{(L)}\in\mathcal{E}^{(L)}} p(\mathbf{y}|\mathbf{x}')^{1/(1+\rho)}\right)^{\rho} d\mathbf{y}, \tag{5.45}
$$

where $\mathbf{x}$ is the signal sequence on $c^{(L)}$ and $\mathbf{x}'$ is the one on $e^{(L)}$.

We can rewrite $q_{LN}(\mathcal{X}) = q_{L(N-1)}(\mathcal{X}'|\mathbf{x})$, where $\mathcal{X}' = \mathcal{X}\backslash\{\mathbf{x}\}$ is the set of signal sequences without $\mathbf{x}$ and $q_{L(N-1)}(\mathcal{X}|\mathbf{x})$ is the probability of $\mathcal{X}\backslash\{\mathbf{x}\}$, conditioned on $\mathbf{x}$. Restricting ρ to the unit interval, $0 \leq \rho \leq 1$, allows us to apply Jensen's inequality, $\sum p_i\alpha_i^{\rho} \leq (\sum p_i\alpha_i)^{\rho}$, and we obtain

$$\text{Avg}\{\overline{P_e}\} \le \sum_{L=v}^{\infty} \frac{1}{2^{kL}} \sum_{c^{(L)}} \int_{\mathbf{y}} \sum_{\mathbf{x}} q(\mathbf{x}) p(\mathbf{y}|\mathbf{x})^{1/(1+\rho)}$$

$$\times \left(\sum_{e^{(L)} \in \mathcal{E}^{(L)}} \underbrace{\sum_{\mathbf{x}_1} \cdots \sum_{\mathbf{x}_N}}_{(\mathbf{x}_i \ne \mathbf{x})} q_{L(N-1)}(\mathcal{X}'|\mathbf{x}) p(\mathbf{y}|\mathbf{x}')^{1/(1+\rho)} \right)^{\rho} d\mathbf{y}. \tag{5.46}$$

But the inner term in (5.46) depends only on $\mathbf{x}'$, and we can reduce (5.46) by summing over all other signal assignment:

$$\text{Avg}\{\overline{P_e}\} \le \sum_{L=v}^{\infty} \frac{1}{2^{kL}} \sum_{c^{(L)}} \int_{\mathbf{y}} \sum_{\mathbf{x}} q(\mathbf{x}) p(\mathbf{y}|\mathbf{x})^{1/(1+\rho)}$$

$$\times \left(\sum_{e^{(L)} \in \mathcal{E}^{(L)}} \sum_{\mathbf{x}'} q(\mathbf{x}'|\mathbf{x}) p(\mathbf{y}|\mathbf{x}')^{1/(1+\rho)} \right)^{\rho} d\mathbf{y}. \tag{5.47}$$

We observe that the last equation depends only on one correct and one incorrect signal sequence. We now further assume that the signals on $c^{(L)}$ and $e^{(L)}$ are assigned independently (i.e., $q(\mathbf{x}'|\mathbf{x}) = q(\mathbf{x}')$), and that each signal is chosen independently, making $q(\mathbf{x}) = \prod_{i=j}^{L-1} q(x_j)$ a product. To make this possible we must assume that the trellis codes are *time varying* in nature, for otherwise each symbol would also depend on the choices of the v most recent symbols. Note also that the signal assignments $\mathbf{x}$ and $\mathbf{x}'$ can be made independently since $e^{(L)}$ is in a different state than $c^{(L)}$ over the entire length of the error event.

This generalization to time-varying codes is quite serious, since almost all practical codes are time invariant. It allows us, however, to obtain a much simpler version of the above bound, starting with

$$\text{Avg}\{\overline{P_e}\} \le \sum_{L=v}^{\infty} \frac{1}{2^{kL}} \sum_{c^{(L)}} \int_{\mathbf{y}} \sum_{\mathbf{x}} \prod_{i=1}^{L} q(x_i) p(y_i|x_i)^{1/(1+\rho)}$$

$$\times \left(\sum_{e^{(L)} \in \mathcal{E}^{(L)}} \sum_{\mathbf{x}'} \prod_{i=1}^{L} q(x_i') p(y_i|x_i')^{1/(1+\rho)} \right)^{\rho} d\mathbf{y}, \tag{5.48}$$

where we have assumed that $p(\mathbf{y}|\mathbf{x}) = \prod_{i=1}^{L} p(y_i|x_i)$—that is, the channel is memoryless. In a final step we realize that the products are now independent of the index j and the correct and incorrect paths $c^{(L)}$ and $e^{(L)}$. We obtain

$$\text{Avg}\left\{\overline{P_e}\right\} \le \sum_{L=v}^{\infty} \int_{\mathbf{y}} \left(\sum_x q(x)p(y|x)^{1/(1+\rho)}\right)^L$$

$$\times \left(N_p \left(\sum_{x'} q(x')p(y|x')^{1/(1+\rho)}\right)^L\right)^{\rho} d\mathbf{y}$$

$$= \sum_{L=v}^{\infty} N_p^{\rho} \int_{\mathbf{y}} \left(\left(\sum_x q(x)p(y|x)^{1/(1+\rho)}\right)^{1+\rho}\right)^L d\mathbf{y}, \tag{5.49}$$

$$\le \sum_{L=v}^{\infty} (2^k - 1)^{\rho} 2^{\rho k(L-v)} \left(\int_y \left(\sum_x q(x)p(y|x)^{1/(1+\rho)}\right)^{1+\rho} dy\right)^L, \tag{5.50}$$

where we have broken up the integral over $\mathbf{y}$ into individual integrals over y in the last step.

Let us now define the error exponent

$$E_0(\rho, \mathbf{q}) \equiv -\log_2 \int_y \left(\sum_x q(x)p(y|x)^{1/(1+\rho)}\right)^{1+\rho} dy, \tag{5.51}$$

where $\mathbf{q} = (q_1, \ldots, q_A)$ is the probability with which x is chosen from a signal set $\mathcal{A}$ of size A. This allows us to write the error bound in the standard form

$$\text{Avg}\left\{\overline{P_e}\right\} \le (2^k - 1)^{\rho} 2^{-vE_0(\rho,\mathbf{q})} \sum_{L=0}^{\infty} 2^{\rho kL} 2^{-LE_0(\rho,\mathbf{q})} \tag{5.52}$$

$$= \frac{(2^k - 1)^{\rho} 2^{-vE_0(\rho,\mathbf{q})}}{1 - 2^{-(E_0(\rho,\mathbf{q})-\rho k)}}, \qquad \rho k < E_0(\rho, \mathbf{q}), \tag{5.53}$$

where we have used the summation formula for a geometric series and the condition $\rho k < E_0(\rho, \mathbf{q})$ ensures convergence.

Since k is the number of information bits transmitted in one channel symbol x, we may call it the information rate in bits per channel use and denote it by R. This gives us our final bound on the event error probability:

$$\text{Avg}\left\{\overline{P_e}\right\} = \frac{(2^R - 1)^{\rho} 2^{-vE_0(\rho,\mathbf{q})}}{1 - 2^{-(E_0(\rho,\mathbf{q})-\rho R)}}, \qquad \rho < \frac{E_0(\rho, \mathbf{q})}{R}. \tag{5.54}$$

Using (5.52) we may easily obtain a bound on the average number of bit errors $\overline{P_b}$. Since on an error path of length L there can be at most $L - v + 1$ bit errors, we obtain

$$\text{Avg}\left\{\overline{P_b}\right\} \le (2^R - 1)^\rho 2^{-\nu E_0(\rho, \mathbf{q})} \sum_{L=0}^{\infty}(L + 1)2^{\rho RL}2^{-LE_0(\rho, \mathbf{q})} \qquad (5.55)$$

$$= \frac{(2^R - 1)^\rho 2^{-\nu E_0(\rho, \mathbf{q})}}{\left(1 - 2^{-(E_0(\rho, \mathbf{q}) - \rho k)}\right)^2}, \qquad \rho < \frac{E_0(\rho, \mathbf{q})}{R}. \qquad (5.56)$$

For large constraint lengths ν, only the exponent $E_0(\rho, \mathbf{q})$ will matter. $E_0(\rho, \mathbf{q})$ is a function of ρ and we wish to explore some of its properties before interpreting our bound. Clearly $E_0(\rho, \mathbf{q}) \ge 0$ for $\rho \ge 0$, since $\sum_x q(x)p(y|x)^{1/(1+\rho)} \le 1$ for $\rho \ge 0$; that is, our exponent is positive, making the bound nontrivial. Furthermore, we have the following.

LEMMA 5.4
$E_0(\rho, \mathbf{q})$ is a monotonically increasing function of ρ.

Proof
We will apply the inequality[5]

$$\left(\sum_i p_i \alpha_i^r\right)^{1/r} \le \left(\sum_i p_i \alpha_i^s\right)^{1/s}, \qquad 0 < r < s, \quad \alpha_i \ge 0, \qquad (5.57)$$

which holds with equality if, for some constant c, $p_i\alpha_i = cp_i$ for all i. Using (5.57) in the expression for the error exponent (5.51), we obtain

$$E_0(\rho, \mathbf{q}) = -\log_2 \int_y \left(\sum_x q(x)p(y|x)^{1/(1+\rho)}\right)^{1+\rho} dy$$

$$\le -\log_2 \int_y \left(\sum_x q(x)p(y|x)^{1/(1+\rho_1)}\right)^{1+\rho_1} dy \qquad (5.58)$$

$$= E_0(\rho_1, \mathbf{q}),$$

for $\rho_1 > \rho > -1$. Equality holds in (5.58) if and only if $p(y|x) = c$ for all $x \in \mathcal{A}$, but that implies that our channel has capacity $C = 0$. Therefore $E_0(\rho, \mathbf{q})$ is monotonically increasing for all interesting cases.

Lemma 5.4 tells us that in order to obtain the best bound, we want to choose ρ as large as possible, given the constraint in (5.54); that is, we choose

---

[5] This inequality is easily derived from Hölder's inequality

$$\sum_i \beta_i \gamma_i \le \left(\sum_i \beta_i^{1/\mu}\right)^\mu \left(\sum_i \gamma_i^{1/(1-\mu)}\right)^{1-\mu}; \qquad 0 < \mu < 1, \quad \beta_i, \gamma_i \ge 0,$$

by letting $\beta_i = p_i^\mu \alpha_i^r$, $\gamma_i = p_i^{1-\mu}$, and $\mu = r/s$.

$$\rho = \frac{E_0(\rho, \mathbf{q})}{R}(1 - \epsilon), \tag{5.59}$$

where ϵ is the usual infinitesimally small number.

Our next lemma says that, unfortunately, as ρ and hence the error exponent increases, the associated rate decreases.

LEMMA 5.5

The function $R(\rho) \equiv E_0(\rho, \mathbf{q})/\rho$ is a monotonically decreasing function of ρ.

Proof

We use the fact that $E_0(\rho, \mathbf{q})$ is a $\cap$-convex function in ρ. This fact is well known and proven in [12; Appendix 3A.3]. We therefore know that $\partial^2/E_0(\rho, \mathbf{q})\partial\rho^2 \leq 0$; that is, the function $\partial/E_0(\rho, \mathbf{q})\partial\rho$ is strictly monotonically decreasing. Let us then consider

$$\frac{\partial}{\partial\rho}\frac{E_0(\rho, \mathbf{q})}{\rho} = \frac{1}{\rho^2}\left(\rho\frac{\partial E_0(\rho, \mathbf{q})}{\partial\rho} - E_0(\rho, \mathbf{q})\right). \tag{5.60}$$

The first term in parentheses lies on a straight line through the origin with slope $\partial E_0(\rho, \mathbf{q})/\partial\rho$. Furthermore, $E_0(\rho, \mathbf{q})$, as a function of ρ, also passes through the origin, and its slope at the origin is larger than $\partial E_0(\rho, \mathbf{q})/\partial\rho$ for all $\rho > 0$. By virtue of the $\cup$-convexity of $E_0(\rho, \mathbf{q})$ the derivative (5.60) is negative, proving the lemma.

We are therefore faced with the situation that the larger the rate R the smaller the maximizing ρ and, hence, the smaller the error exponent $E_0(\rho, \mathbf{q})$. In fact, $E_0(\rho, \mathbf{q}) \to 0$ as $\rho \to 0$, which can easily be seen from (5.51).

Let us find out at what limiting rate R the error exponent approaches its value of zero; that is, let us calculate

$$\lim_{\rho\to 0} R(\rho) = \lim_{\rho\to 0} \frac{E_0(\rho, \mathbf{q})}{\rho}. \tag{5.61}$$

Since this limit is of the form $0/0$, we can use l'Hôpital's rule and obtain the limit

$$\lim_{\rho\to 0} R(\rho) = \left.\frac{\partial E_0(\rho, \mathbf{q})}{\partial\rho}\right|_{\rho=0}. \tag{5.62}$$

This derivative is easily evaluated, and to our great surprise and satisfaction we obtain

$$\left.\frac{\partial E_0(\rho, \mathbf{q})}{\partial \rho}\right|_{\rho=0} = \int_y \sum_x q(x)p(y|x) \log_2 \left(\frac{p(y|x)}{\sum_{x'} q(x')p(y|x')}\right) dy \tag{5.63}$$

$$= C(\mathbf{q}),$$

the Shannon channel capacity using the signal set $\mathcal{A}$ and the input probability distribution $\mathbf{q}$! (See also [11, 12].) Maximizing over the input probability distribution $\mathbf{q}$ will then achieve the channel capacity C.

We have now proved the important fact that group-trellis codes can achieve the Shannon capacity, and, as we will see, with a very favorable error exponent (5.51). The bounds thus derived are an important confirmation of our decision to use group-trellis codes, since they tell us that these codes can achieve channel capacity.

Let us now concern ourselves with additive white Gaussian noise (AWGN) channels, since they are arguably the most important class of memoryless channels. For an AWGN channel with complex input and output symbols x and y (from (2.11))

$$p(y|x) = \frac{1}{2\pi N_0} \exp\left(-\frac{|y-x|^2}{2N_0}\right). \tag{5.64}$$

Furthermore, for $\rho = 1$, $E_0(\rho, \mathbf{q})$ can be simplified considerably by evaluating the integral over v:

$$E_0(1, \mathbf{q}) = -\log_2 \int_y \left(\sum_x q(x) \frac{1}{\sqrt{2\pi N_0}} \exp\left(-\frac{|y-x|^2}{4N_0}\right)\right)^2$$

$$= -\log_2 \sum_x \sum_{x'} q(x)q(x') \int_y \frac{1}{2\pi N_0} \exp\left(-\frac{|y-x|^2}{4N_0} - \frac{|y-x'|^2}{4N_0}\right)$$

$$= -\log_2 \sum_x \sum_{x'} q(x)q(x') \exp\left(-\frac{|x-x'|^2}{4N_0}\right) = R_0(\mathbf{q}), \tag{5.65}$$

where $R_0(\mathbf{q}) = E_0(1, \mathbf{q})$ is known in the literature as the cutoff rate of the channel. R_0 is believed by many experts [13, 14] to be the "practical" limit for coded systems. Indeed, there are coding schemes that achieve R_0. Furthermore, in Chapter 6 we will see that certain decoding algorithms exhibit an unbounded computation time for rates that exceed R_0.

On AWGN channels the uniform input symbol distribution $\mathbf{q} = 1/A, \ldots, 1/A$ is particularly popular, and we define $E_0(\rho, \text{uniform}) = E_0(\rho)$ and $R_0(\text{uniform}) = R_0$; that is, we simply omit the probability distribution vector $\mathbf{q}$.

Figure 5.10 shows $E_0(\rho)$ for an 8-PSK constellation on an AWGN channel with signal-to-noise ratio $E_b/N_0 = 10$ dB. The error exponent

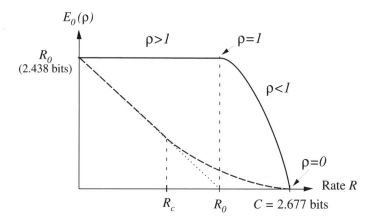

Figure 5.10 Error exponent as a function of the rate R for an 8-PSK constellation on an AWGN channel at a signal-to-noise ratio of $E_b/N_0 = 10$ dB.

function exhibits the typical behavior, known from error bounds for convolutional codes [12], having a constant value of R_0 up to the rate $R = R_0$ and then rapidly dropping to zero as $R \to C$. We have also shown the error exponent for block codes[6] (dashed curve) in Figure 5.10. A thorough discussion about block code error exponents can be found in [12] or [15].

We discern the interesting fact that the error exponent for trellis codes is significantly larger than that for block codes, especially at high rates of transmission. Since the block code error exponent multiplies the block code length, this is taken as an indication that the constraint length ν needed to obtain a prescribed error probability $\overline{P_b}$ is much smaller than the block length of an equivalent block code. This may account for the popularity of trellis codes and their often superior behavior on high-noise channels. Although most academic publications are devoted to block codes, trellis (particularly convolutional) codes are more widespread in applications.

Figure 5.11 shows the capacity C and the cutoff rate R_0 as functions of the symbol-to-noise ratio E_S/N_0 for PSK constellations. The performance points in the figure refer to uncoded transmission, and the possible gains that coding can provide become visible. Such capacity curves were originally used to motivate the use of expanded signal sets in trellis coding by Ungerböck in his original paper [16].

---

[6] The block error probability of the ensemble of block codes of length N is given by [12, Section 3.1]

$$\overline{P_B} < 2^{-NE(R)}, \qquad E(R) = \max_{\mathbf{q}} \max_{0 \le \rho \le 1} \left[E_0(\rho, \mathbf{q}) - \rho R \right].$$

C, R_0 [bits/symbol]

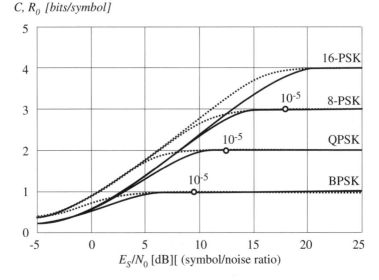

Figure 5.11 Capacity and cutoff rate as a function of the symbol-to-noise ratio for selected PSK constellations.

In this last section we discussed some basic information-theoretic concepts as they relate to trellis codes. To probe further, the books by Gallager [15] and Viterbi/Omura [12] are classics and good starting points.

APPENDIX A

The following theorem is also known as Gerschgorin's *circle theorem*.

THEOREM 5.6

Every eigenvalue λ_i of a matrix $\mathbf{P}$ with arbitrary complex entries lies in at least one of the circles C_i, whose centers are at p_{ii} and whose radii are $r_i = \sum_{j \neq i} |p_{ij}|$; i.e., r_i is the ith absolute row sum without the diagonal element.

Proof

$\mathbf{P}\mathbf{x} = \lambda\mathbf{x}$ immediately leads to

$$(\lambda - p_{ii})x_i = \sum_{j \neq i} p_{ij}x_j. \tag{5.66}$$

Taking absolute values on both sides and applying the triangle inequality, we obtain

$$|\lambda - p_{ii}| \leq \sum_{j \neq i} |p_{ij}||x_j|/|x_i|. \tag{5.67}$$

Now let x_i be the largest component of $\mathbf{x}$. Then $|x_j|/|x_i| \leq 1$, and $|\lambda - p_{ii}| \leq r_i$.

In the application in Section 5.3, all entries $p_{ij} \geq 0$, and hence

$$(\lambda - p_{ii}) \leq \sum_{j \neq i} p_{ij} \Rightarrow \lambda \leq \sum_j p_{ij}. \tag{5.68}$$

Now, the largest eigenvalue $\lambda_{\max}$ must be smaller than the largest row sum:

$$\lambda_{\max} \leq \max_i \sum_j p_{ij}. \tag{5.69}$$

APPENDIX B

In this appendix we describe an efficient algorithm to produce an equivalent FSM to a given FSM $\mathcal{M}$, which has the minimal number of states among all equivalent FSMs. The algorithm starts out by assuming all states are equivalent, and then it successively partitions the sets of equivalent states until all true equivalent states are found. The algorithm terminates in a finite number of steps, since this refinement of the partitioning must end when each original state is in an equivalent set by itself. The algorithm performs the following steps:

Step 1.   Form a first partition P_1 of the states of $\mathcal{M}$ by grouping states that produce identical sets of outputs $\delta(u \rightarrow u')$ as we go through all transitions $u \rightarrow u'$ (compare Definition 5.2).

Step 2.   Obtain the $(l + 1)$th partition P_{l+1} from the lth partition P_l as follows: two states u and v are in the same equivalent group of P_{l+1} if and only if

   i u and v are in the same equivalent group of P_l, and

   ii for each pair of transitions $u \rightarrow u'$ and $v \rightarrow v'$ that produce the same output, u' and v' are in the same equivalent set of P_l.

Step 3.   Repeat Step 2 until no further refinement occurs, that is, until $P_m = P_{m-1}$. P_m is the final desired partition of the states of $\mathcal{M}$ into equivalent states. Any member of the group can now be used to represent the entire group.

The recursive nature of the algorithm quickly proves that it actually produces the desired partition. Namely, if two states are in P_l, all their *sequences* of output sets must be identical for l steps, since their successors are in P_{l-1}, etc., all the way back to P_1, which is the first and largest group of states that produce identical sets of outputs.

Now, if $P_{l-1} = P_l$, the successors of the above two states are in P_l also, so their sequences of output sets are identical for l steps also. We conclude that their sequences of output sets are therefore identical for all time.

The algorithm can be extended to input and output equivalence of FSMs. The interested reader will find a more complete discussion in [17, Chapter 4].

REFERENCES

[1] W. Zhang, "Finite-state systems in mobile communications," Ph.D. Dissertation, University of South Australia, June 1995.

[2] E. Biglieri, "High-level modulation and coding for nonlinear satellite channels," *IEEE Trans. Commun.*, Vol. COM-32, pp. 616–626, 1984.

[3] J. M. Wozencraft and I. M. Jacobs, *Principles of Communiation Engineering*, Wiley, New York, 1965.

[4] F. L. Alvarado, "Manipulation and visualization of sparse matrices," *ORSA J. Comput.*, Vol. 2, No. 2, 1990, pp. 186–207.

[5] G. Strang, *Linear Algebra and Its Applications*, Harcourt Brace Jovanovich, San Diego, 1988.

[6] E. Zehavi and J. K. Wolf, "On the performance evaluation of trellis codes," *IEEE Trans. Inform. Theory*, Vol. IT-33, No. 2, pp. 196–201, 1987.

[7] M. Rouanne and D. J. Costello, Jr., "An algorithm for computing the distance spectrum of trellis codes," *IEEE J. Select. Areas Commun.*, Vol. SAC-7, No. 6, pp. 929–940, 1989.

[8] C. Schlegel, "Evaluating distance spectra and performance bounds of trellis codes on channels with intersymbol interference," *IEEE Trans. Inform. Theory*, Vol. IT-37, No. 3, pp. 627–634, 1991.

[9] E. Biglieri and P. J. McLane, "Uniform distance and error probability properties of TCM schemes," *IEEE Trans. Commun.*, Vol. COM-39, pp. 41–53, 1991.

[10] W. Zhang and C. Schlegel, "State reduction in the computation of d_{free} and the distance spectrum for a class of convolutional codes," *Proc. SICON/ICIE 93, Singapore*, September 1993.

[11] T. M. Cover and J. A. Thomas, *Elements of Information Theory*, Wiley, New York, 1991.

[12] A. J. Viterbi and J. K. Omura, *Principles of Digital Communication and Coding*, McGraw-Hill, New York, 1979.

[13] J. M. Wozencraft and R. S. Kennedy, "Modulation and demodulation for probabilistic coding," *IEEE Trans. Inform. Theory*, Vol. IT-12, No. 3, pp. 291–297, 1966.

[14] J. L. Massey, "Coding and modulation in digital communications," *Proc. Int. Zürich Sem. Digital Commun.*, Zürich, Switzerland, March 1974, pp. E2(1)–E2(4).

[15] R. G. Gallager, *Information Theory and Reliable Communication*, Wiley, New York, 1968.

[16] G. Ungerböck, "Channel coding with multilevel/phase signals," *IEEE Trans. Inform. Theory*, Vol. IT-28, No. 1, pp. 55–67, 1982.

[17] P. J. Denning et al., *Machines, Language and Computation*, Prentice Hall, Englewood Cliffs, NJ, 1978.

CHAPTER 6

DECODING STRATEGIES

6.1 TREE DECODERS

In Chapters 2 and 3 we discussed how a trellis encoder generates a sequence $\mathbf{x}^{(i)} = (x_{-l}^{(i)}, \ldots, x_l^{(i)})$ of correlated complex symbols $x_r^{(i)}$ for message i and how this sequence is modulated, using the pulse waveform $p(t)$, into the (baseband) output signal

$$s^{(i)}(t) = \sum_{r=-l}^{l} x_r^{(i)} p(t - rT). \tag{6.1}$$

We have assumed in (6.1) that two-dimensional signal constellations are used. If our trellis code generates multidimensional output symbols, these are modulated in blocks of two-dimensional symbols.

From Chapter 2 we also know the structure of the optimal decoder for such a system. We have to build a matched filter for each possible signal $s^{(i)}(t)$ and select the message that corresponds to the signal that produces the largest sampled output value.

The matched filter for $s^{(i)}(t)$ is given by

$$s^{(i)}(-t) = \sum_{r=-l}^{l} x_r^{(i)} p(-t - rT), \tag{6.2}$$

and, if $r(t)$ is the received signal, the sampled response of the matched filter (6.2) to $r(t)$ is given by (see also (2.21))

$$\mathbf{r} \cdot \mathbf{s}^{(i)} = \int_{-\infty}^{\infty} r(t) s^{(i)}(t) \, dt$$

$$= \sum_{r=-l}^{l} x_r^{(i)} y_r = \mathbf{x}^{(i)} \cdot \mathbf{y}, \tag{6.3}$$

where $y_r = \int_{-\infty}^{\infty} r(\alpha)p(\alpha - rT)$ is the output of the filter matched to the pulse $p(t)$ sampled at time $t = rT$, as discussed in Section 2.5 (equation (2.24)), and $\mathbf{y} = (y_{-l}, \ldots, y_l)$ is the vector of sampled signals y_r.

If time-orthogonal pulses (e.g., Nyquist pulses) $p(t)$ with unit energy $(\int_{-\infty}^{\infty} p^2(t)\, dt = 1)$ are used, the energy of the signal $s^{(i)}(t)$ is

$$|\mathbf{s}^{(i)}|^2 = \int_{-\infty}^{\infty}\int_{-\infty}^{\infty} s^{(i)}(\alpha)s^{(i)}(\beta)\, d\alpha\, d\beta$$

$$= \sum_{r=-l}^{l} |x_r^{(i)}|^2, \tag{6.4}$$

and, from (2.12), the maximum likelihood receiver will select the sequence $\mathbf{x}^{(i)}$ that maximizes

$$J^{(i)} = 2\sum_{r=-l}^{l} \text{Re}\left\{x_r^{(i)} y_r^*\right\} - \sum_{r=-l}^{l} |x_r^{(i)}|^2. \tag{6.5}$$

Equation (6.5) is called the *metric* of the sequence $\mathbf{x}^{(i)}$, and this metric is to be maximized over all allowable choices of $\mathbf{x}^{(i)}$. We may now define the partial metric at time n as

$$J_n^{(i)} = 2\sum_{r=-l}^{n} \text{Re}\left\{x_r^{(i)} y_r^*\right\} - \sum_{r=-l}^{n} |x_r^{(i)}|^2, \tag{6.6}$$

which allows us to rewrite (6.5) in the recursive form

$$J_n^{(i)} = J_{n-1}^{(i)} + 2\text{Re}\left\{x_n^{(i)} y_n^*\right\} - |x_n^{(i)}|^2. \tag{6.7}$$

Equation (6.7) implies a tree structure to evaluate the metrics for all the allowable signal sequences as illustrated in Figure 6.1 for the trellis code from Figure 3.1. The tree has, in general, 2^k branches leaving each node, since there are 2^k possible different choices of the signal x_n at time n. Each node is labeled with the hypothesized partial sequence[1] $\tilde{\mathbf{x}}^{(i)} = (x_{-l}^{(i)}, \ldots, x_n^{(i)})$ that leads to it. The intermediate metric $J_n^{(i)}$ is also stored for each node. The tree starts at time $n = -l$ and extends until time unit $n = l$, at which time the largest accumulated metric identifies the most likely sequence of symbols $\mathbf{x}^{(i)}$.

It becomes obvious that this tree grows very quickly. In fact, its final size is k^{2l+1}, which is an outlandish number even for small values of l (i.e., short encoded sequences). We therefore need to reduce the complexity of

---

[1] We denote partial sequences by tildes to distinguish them from complete sequences or codewords.

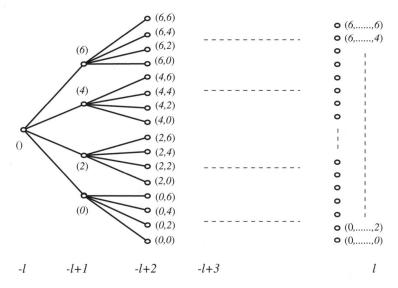

Figure 6.1 Metric tree for a trellis decoder extending from time $-l$ to time l for the code from Figure 3.1.

decoding in some appropriate way, and this can be done by performing only a partial search of the tree.

There are different approaches to tree decoding, and we discuss the more fundamental types in the subsequent sections. Before we tackle these decoding algorithms, however, we wish to modify the metric such that it takes into account the different lengths of paths, since we will come up against the problem of comparing paths of different lengths in the tree.

Consider the set $\mathcal{X}_M$ of M partial sequences $\tilde{\mathbf{x}}^{(i)}$ of length n_i, and let $n_{\max} = \max\{n_1, \dots, n_M\}$ be the maximum length among the M partial sequences. The decoder must make its likelihood ranking of the paths based on the partial received sequence $\tilde{\mathbf{y}}$ of length $n_{\max}$.

From (2.10) we know that an optimal receiver chooses the $\tilde{\mathbf{x}}^{(i)}$ that maximizes

$$P[\tilde{\mathbf{x}}^{(i)}|\tilde{\mathbf{y}}] = P[\tilde{\mathbf{x}}^{(i)}]\frac{\prod_{r=-l}^{-l+n_i} p_n(y_r - x_r) \prod_{r=-l+n_i}^{-l+n_{\max}} p(y_r)}{p(\tilde{\mathbf{y}})}, \qquad (6.8)$$

where the second product reflects the fact that we have no hypotheses $x_r^{(i)}$ for $r > n_i$, since $\tilde{\mathbf{x}}^{(i)}$ extends only from $-l$ to $-l + n_i$. We therefore have to use the a priori probabilities $p(y_r|x_r^{(i)}) = p(y_r)$ for $r > n_i$. Using $p(\tilde{\mathbf{y}}) = \prod_{r=-l}^{-l+n_{\max}} p(y_r)$, we rewrite equation (6.8) as

$$P[\tilde{\mathbf{x}}^{(i)}|\tilde{\mathbf{y}}] = P[\tilde{\mathbf{x}}^{(i)}] \prod_{r=-l}^{-l+n_i} \frac{p_n(y_r - x_r^{(i)})}{p(y_r)}, \tag{6.9}$$

and we see that we need not be concerned with the tail samples not affected by $\tilde{\mathbf{x}}_i$.

We may take logarithms now to obtain an additive metric, given by

$$L(\tilde{\mathbf{x}}^{(i)}, \tilde{\mathbf{y}}) = \sum_{r=-l}^{-l+n_i} \log \frac{p_n(y_r - x_r^{(i)})}{p(y_r)} - \log \frac{1}{P[\tilde{\mathbf{x}}^{(i)}]}. \tag{6.10}$$

Since $P[\tilde{\mathbf{x}}^{(i)}] = (2^{-k})^{n_i}$ is the probability of the partial sequence $\tilde{\mathbf{x}}^{(i)}$, assuming that all the inputs to the trellis encoder have equal probability, (6.10) becomes

$$L(\tilde{\mathbf{x}}^{(i)}, \tilde{\mathbf{y}}) = L(\tilde{\mathbf{x}}^{(i)}, \mathbf{y}) = \sum_{r=-l}^{-l+n_i} \left[\log \frac{p_n(y_r - x_r^{(i)})}{p(y_r)} - k \right], \tag{6.11}$$

where we have extended $\tilde{\mathbf{y}} \to \mathbf{y}$ since (6.11) ignores the tail samples $y_r, r > n_i$, anyway. The metric (6.11) was introduced for decoding tree codes by Fano [1] in 1963, and it was analytically derived by Massey [2] in 1972.

Since equation (2.11) explicitly gives the conditional probability distribution $p_n(y_r - x_r)$, the metric in (6.11) can be specialized for additive white Gaussian noise channels to

$$\begin{aligned} L(\tilde{\mathbf{x}}^{(i)}, \mathbf{y}) &= \sum_{r=-l}^{-l+n_i} \left[\log \frac{\exp\left(-|x_r^{(i)} - y_r|^2/N_0\right)}{\sum_{x \in A} p(x) \exp\left(-|x - y_r|^2/N_0\right)} - k \right] \\ &= -\sum_{r=-l}^{-l+n_i} \frac{|x_r^{(i)} - y_r|^2}{N_0 - c_r(y_r)}, \end{aligned} \tag{6.12}$$

where $c_r(y_r) = \log\left(\sum_{x \in A} p(x) \exp\left(-|x - y_r|^2/N_0\right)\right) + k$ is a term independent of $x_r^{(i)}$ that is subtracted from all the metrics at time r. Note that c_r can be positive or negative, which is one of the problems with sequential decoding.

If the paths examined are of the same length, say n, they all contain the same cumulative constant $-\sum_{r=-l}^{-l+n} c_r$ in their metrics, which therefore may be discarded from all the metrics. This allows us to simplify (6.12) to

$$L(\tilde{\mathbf{x}}^{(i)}, \mathbf{y}) \equiv \sum_{r=-l}^{-l+n} 2\mathrm{Re}\left\{x_r^{(i)} y_r^*\right\} - |x_r^{(i)}|^2 = J_n^{(i)}, \tag{6.13}$$

by neglecting terms common to all the metrics. The metric (6.13) is equivalent to the accumulated Euclidean distance between the received partial sequence $\tilde{\mathbf{y}}$ and the hypothesized symbols on the ith path up to length n. The restriction to paths of equal length makes this metric much simpler than the general metric (6.11) (and (6.12)) and finds application in the *breadth-first* decoding algorithms, which we discuss in subsequent sections.

6.2   THE STACK ALGORITHM

The stack algorithm is one of the many variants of what has become known as sequential decoding of trellis codes. Sequential decoding was introduced by Wozencraft [3] for convolutional codes and has subsequently experienced many changes and additions. Sequential decoding describes any algorithm for decoding trellis codes that successively explores the encoder tree by moving to new nodes from an already explored node.

From the introductory discussion in the preceding section, one way of sequential decoding becomes apparent. We start exploring the tree and store the metric (6.11) (or (6.12)) for every node explored. At each stage now we simply extend the node with the largest such metric. This, in essence, is the *stack algorithm* first proposed by Zigangirov [4] and Jelinek [5]. This basic algorithm is then:

Step 1.   Initialize an empty stack S of visited nodes and their metrics. Deposit the empty partial sequence ( ) at the top of the stack with its metric $L(( ), \mathbf{y}) = 0$.

Step 2.   Extend the node corresponding to the top entry $\left\{ \tilde{\mathbf{x}}_{\text{top}}, L(\tilde{\mathbf{x}}_{\text{top}}, \mathbf{y}) \right\}$ by forming $L(\tilde{\mathbf{x}}_{\text{top}}, \mathbf{y}) - |x_r - y_r|^2/N_0 - c_r$ for all 2^k extensions of $\tilde{\mathbf{x}}_{\text{top}} \rightarrow (\tilde{\mathbf{x}}_{\text{top}}, x_r) = \tilde{\mathbf{x}}^{(i)}$. Delete $\left\{ \tilde{\mathbf{x}}_{\text{top}}, L(\tilde{\mathbf{x}}_{\text{top}}, \mathbf{y}) \right\}$ from the stack.

Step 3.   Place the new entries $\left\{ \tilde{\mathbf{x}}^{(i)}, L(\tilde{\mathbf{x}}^{(i)}, \mathbf{y}) \right\}$ from Step 2 into the stack such that the stack remains ordered with the entry with the largest metric at the top of the stack.

Step 4.   If the top entry of the stack is a path to one of the terminal nodes at depth l, stop and select $\mathbf{x}_{\text{top}}$ as the transmitted symbol sequence. Otherwise, go to Step 1.

There are some practical problems associated with the stack algorithm. First, the number of computations that the algorithms perform depends greatly on channel quality. If we have a very noisy channel, the received sample value y_r will be unreliable and numerous possible paths will have similar metrics. These paths all have to be stored in the stack and further explored. This causes a computational speed problem, since the incoming symbols have to be stored in a buffer while the algorithm performs the decoding operation. This buffer is now likely to overflow if the channel is very noisy and the decoder will have to declare a decoding failure. This phenomenon is explored further in Section 6.7. In practice, the transmitted data will be framed and the decoder will declare a frame erasure if it experiences input buffer overflow.

A second problem with the stack algorithm is the increasing complexity of Step 2 (i.e., of reordering the stack). This sorting operation depends on the size of the stack, which, again, for very noisy channels becomes large. This problem is addressed in all practical applications by ignoring small differences in the metric and collecting all stack entries with metrics within a specified "quantization interval" in the same bucket. Bucket j contains all stack entries with metrics

$$j\Delta \leq L(\tilde{\mathbf{x}}^{(i)}, \mathbf{y}) \leq (j + 1)\Delta, \qquad (6.14)$$

where Δ is a variable quantization parameter. Incoming paths are now sorted only into the correct bucket, avoiding the sorting complexity of the large stack. The depositing and removal of the paths from the buckets can occur on a "last in, first out" basis.

There are a number of variations of this basic theme. If Δ is a fixed value, the number of buckets can also grow to be large, and the sorting problem, originally avoided, reappears. An alternative is to let the buckets vary in size rather than in metric range. In that way, the critical dependence on the stack size can be avoided.

An associated problem with the stack is that of stack overflow. This is less severe and the remedy is simply to drop the last path in the stack from future consideration. The probability of actually losing the correct path is very small, much smaller than the problem of a frame erasure. A large number of variants of this algorithm are feasible and have been explored in the literature. Further discussion of the details of implementation of these algorithms are found in [6–12].

6.3   THE FANO ALGORITHM

Unlike the stack algorithm, the Fano algorithm is a depth-first tree search procedure in its purest form. Introduced by Fano [1] in 1963, this algorithm stores only one path and thus, essentially, requires no storage. Its drawback is a certain loss in speed compared to the stack algorithm for higher rates [13], but for moderate rates the Fano algorithm decodes faster than the stack algorithm [14]. It seems that the Fano algorithm is the preferred choice for practical implementations of sequential decoding algorithms.

Since the Fano algorithm only stores one path, it must allow for back-tracking. Also, there can be no jumping between nonconnected nodes; that is, the algorithm only moves between adjacent nodes that are connected in the code tree. The algorithm starts at the initial node and moves in the tree by proceeding from one node to a successor node with a suitably large metric. If no such node can be found, the algorithm backtracks and looks for other branches leading off from previously visited nodes. The metrics of all these adjacent nodes can be computed by adding or subtracting the metric of the connecting branch, and no costly storing of metrics is re-quired. If a node is visited more than once, its metric is recomputed. This is part of the computation/storage trade-off of sequential decoding.

The algorithm proceeds along a chosen path as long as the metric con-tinues to increase. It does that by continually tightening a metric threshold to the current node metric as it visits nodes for the first time. If new nodes along the path have a metric smaller than the threshold, the algorithm backs up and looks at other node extensions. If no other extensions with a metric above the threshold can be found, the value of the threshold is decreased and the forward search is resumed. In this fashion each node visited in the forward direction more than once is reached with a progressively lower threshold each time. This prevents the algorithm from getting caught in an infinite loop. Eventually this procedure reaches a terminal node at the end of the tree and a decoded symbol sequence can be output.

Figure 6.2 depicts an example of the search behavior of the Fano algo-rithm. Assume that there are two competing paths, where the solid path is the most likely sequence and the dashed path is a competitor. The vertical height of the nodes in Figure 6.2 is used to illustrate the metrics of each of the nodes. Also assume that the paths shown are those with the best metrics; that is, all other branches leading off from the nodes lead to nodes with smaller metrics. Initially, the algorithm will proceed to node A, at which time it will start to backtrack since the metric of node D is smaller than that of node A. After exploring alternatives and successively lowering the threshold to t_1 and then to t_2, it will reach node O again and proceed along the dashed path to node B and node C. Now it will start to backtrack again, lowering its threshold to t_3 and then to t_4. It will now again explore the solid path beyond node D to node E, since the lower threshold will allow that. From there the path metrics pick up again and the algorithm proceeds along the solid path. If the threshold decrement Δ had been twice as large, the algorithm would have moved back to node O faster, but would

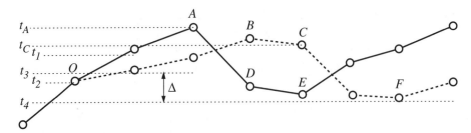

Figure 6.2 Illustration of the operation of the Fano algorithm when choosing between two competing paths.

also have been able to move beyond the metric dip at node F, and would have chosen the erroneous path.

It becomes obvious that at some point the metric threshold t will have to be lowered to the lowest metric value that the maximum likelihood path assumes, and, consequently, a large decrement Δ allows the decoder to achieve this low threshold faster. Conversely, if the decrement Δ is too large, t may drop to a value that allows several erroneous paths to be potentially decoded before the maximum metric path. The optimal value of the metric threshold is best determined by experience and simulations.

Figure 6.3 shows the flowchart of the Fano algorithm.

6.4   THE M-ALGORITHM

This section deals with a purely breadth-first algorithm. The M-algorithm is a synchronous algorithm that moves from time unit to time unit. It keeps M candidate paths at each iteration and deletes all others from further consideration. At each time unit the algorithm extends all M currently held nodes to form $2^k M$ new nodes, from among which those M with the best metrics are retained. Due to the breadth-first nature of the algorithm, the metric in (6.13) can be used. The algorithm is very simply:

Step 1.   Initialize an empty list L of candidate paths and their metrics. Deposit the zero-length path ( ) with its metric $L(( ), \mathbf{y}) = 0$ in the list. Set $n = -l$.

Step 2.   Extend 2^k partial paths $\tilde{\mathbf{x}}^{(i)} \rightarrow (\tilde{\mathbf{x}}^{(i)}, x_n^{(i)})$ from each of the at-most M paths $\tilde{\mathbf{x}}^{(i)}$ in the list. Delete the entries in the original list.

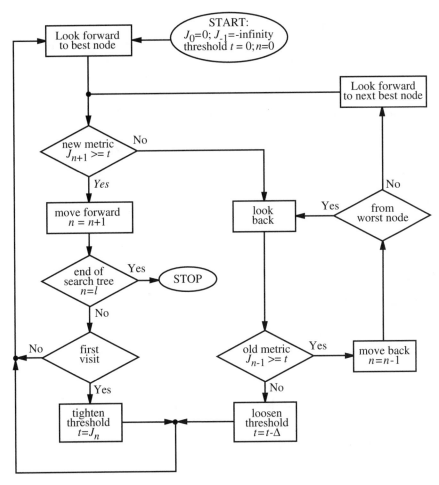

Figure 6.3 Flowchart of the Fano algorithm. The initialization of $J_{-1} = -\infty$
has the effect that the algorithm can lower the threshold for the first
step, if necessary.

Step 3.   Find the at-most M partial paths with the best metrics among
the extensions[2] and save them in the list L.   Delete the rest of the
extensions.   Set $n = n + 1$.

Step 4.   If at the end of the tree (i.e., $n > l$) release the output symbols
corresponding to the path with the best metric in the list L; otherwise
go to Step 2.

---

[2] Note that from two or more extensions leading to the same state (see Section 6.5) all but the one
with the best metric may be discarded. This will improve performance slightly by eliminating some
paths which cannot be the maximum-likelihood (ML) path.

This algorithm is straightforward to implement and its popularity is partly due to the simple metric as compared to sequential decoding. The decoding problem with the M-algorithm is the loss of the correct path from the list of candidates, after which the algorithm might spend a long time resynchronizing. This problem is usually addressed by framing the data. With each new frame resynchronization is achieved. Another option is to use the M-algorithm with block codes (see Section 7.7). The computational load of the M-algorithm is independent of the size of the code; it is proportional to M. Unlike depth-first algorithms, it is also independent of the quality of the channel, since M paths are retained irrespective of the channel quality.

A variant of the M-algorithm is the T-algorithm. It differs from the M-algorithm only in Step 3, where instead of a fixed number M, all paths with metrics $L(\tilde{\mathbf{x}}^{(i)}, \mathbf{y}) \geq \lambda_t - T$ are retained, where λ_t is the metric of the best path and T is some arbitrary threshold. The T-algorithm is therefore in a sense a hybrid between the M-algorithm and a stack-type algorithm. Its performance depends on T, but is very similar to that of the M-algorithm, and we will not discuss it further.

In Chapter 3 we have discussed that the performance of trellis codes using an ML detector was governed by the distance spectrum of the code, where the minimum free-squared Euclidean distance d_{free} played a particularly important role. Since the M-algorithm is a suboptimal decoding algorithm, its performance is additionally affected by other criteria. The major criterion is the probability that the correct path is not among the M retained candidates at time n. If this happens, we lose the correct path and it usually takes a long time to resynchronize. We will see that the probability of correct path loss has no direct connection to the distance spectrum of a code.

The complexity of the M-algorithm resides in the M path extensions at each stage and is largely independent of the code size and constraint length. For this reason, one usually chooses very long constraint-length codes to ensure that the minimum free-squared Euclidean distance is appropriately large. If this is the case, the correct path loss becomes the dominant error event of the decoder.

Let us then take a closer look at the probability of losing the correct path at time n. We assume that at time $n - 1$ the correct path was among the M retained candidates, as illustrated in Figure 6.4. Each of these M nodes is extended into 2^k nodes at time n, of which M are to be retained. There are then a total of $\binom{M2^k}{M}$ ways of choosing the new M retained paths at time n.

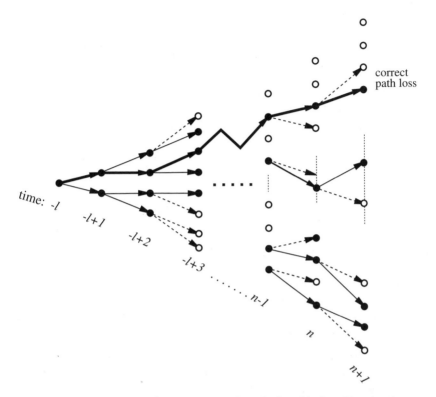

Figure 6.4 Extension of $2^k M = 2 \cdot 4$ paths from the list of the best $M = 4$ paths. The solid paths are those retained by the algorithm; the path indicated by the heavy line corresponds to the correct transmitted sequence.

Let us denote the correct partial path by $\tilde{\mathbf{x}}^{(c)}$. The optimal strategy of the decoder will then be to retain that particular set of M candidates that maximizes the probability of containing $\tilde{\mathbf{x}}^{(c)}$. Let C_p be one of the $\binom{M2^k}{M}$ possible sets of M candidates at time n. Then we wish to maximize

$$\max_p \Pr\left\{\tilde{\mathbf{x}}^{(c)} \in C_p | \tilde{\mathbf{y}}\right\}. \tag{6.15}$$

Since all the partial paths $\tilde{\mathbf{x}}^{(p_j)} \in C_p$, $j = 1, \ldots, M$, are distinct, the events $\{\tilde{\mathbf{x}}^{(c)} = \tilde{\mathbf{x}}^{(p_j)}\}$ are all mutually exclusive for different j; that is, the correct path can be at most only one of the M different candidates $\tilde{\mathbf{x}}^{(p_j)}$. Equation (6.15) can therefore be evaluated as

$$\max_p \Pr\left\{\tilde{\mathbf{x}}^{(c)} \in C_p | \tilde{\mathbf{y}}\right\} = \max_p \sum_{j=1}^{M} \Pr\left\{\tilde{\mathbf{x}}^{(c)} = \tilde{\mathbf{x}}_j^{(p_j)} | \tilde{\mathbf{y}}\right\}. \tag{6.16}$$

From (6.8), (6.10), and (6.13) we know that

$$\Pr\left\{\tilde{\mathbf{x}}^{(c)} = \tilde{\mathbf{x}}^{(p_j)}|\tilde{\mathbf{y}}\right\} \propto \exp\left(-\sum_{r=-l}^{-l+n}\left(2\mathrm{Re}\left\{x_r^{(p_j)}y_r^*\right\} - |x_r^{(p_j)}|^2\right)\right), \quad (6.17)$$

where the proportionality constant is independent of $\tilde{\mathbf{x}}^{(p_j)}$. The maximization in (6.15) now becomes equivalent to (considering only the exponent from above)

$$\max_p \Pr\left\{\tilde{\mathbf{x}}^{(c)} \in \mathcal{C}_p|\tilde{\mathbf{y}}\right\} \equiv \max_p \sum_{j=1}^M \sum_{r=-l}^{-l+n}\left(2\mathrm{Re}\left\{x_r^{(p_j)}y_r^*\right\} - |x_r^{(p_j)}|^2\right)$$

$$= \max_p J_n^{(p_j)}; \qquad (6.18)$$

that is, we simply collect the M paths with the best partial metrics $J_n^{(p_j)}$ at time n. This was shown by Aulin [15]. Earlier we showed that the total metric can be broken up into the recursive form of (6.7), but now we have shown that if the detector is constrained to considering only a maximum of M paths at each stage, retaining those M paths $\tilde{\mathbf{x}}^{(p_j)}$ with maximum partial metrics is the optimal strategy.

The probability of correct path loss, denoted by Pr(CPL), can now be addressed. Follow the methodology of Aulin [15], we need to evaluate the probability that the correct path $\tilde{\mathbf{x}}^{(c)}$ is not among the M candidate paths. This will happen if M paths $\tilde{\mathbf{x}}^{(p_j)} \neq \tilde{\mathbf{x}}^{(c)}$ have a partial metric $J_n^{(p_j)} \geq J_n^{(c)}$ or, equivalently, if all the M metric differences

$$\delta_n^{(j,c)} = J_n^{(c)} - J_n^{(p_j)} = \sum_{r=-l}^{-l+n}\left(|x_r^{(p_j)}|^2 - |x_r^{(c)}|^2 - 2\mathrm{Re}\left\{x_r^{(p_j)} - x_r^{(c)}\right\}y_r^*\right) \quad (6.19)$$

are smaller than or equal to zero. That is,

$$\Pr(\mathrm{CPL}|\mathcal{C}_p) = \Pr\{\delta_n \leq \mathbf{0}\}, \qquad \tilde{\mathbf{x}}^{(p_j)} \in \mathcal{C}_p, \qquad (6.20)$$

where $\delta_n = \left(\delta_n^{(1,c)}, \ldots, \delta_n^{(M,c)}\right)$ is the vector of metric differences at time n. $\Pr(\mathrm{CPL}|\mathcal{C}_p)$ is the probability of correct path loss for a specific set $\mathcal{C}_p$ of M retained paths not containing $\tilde{\mathbf{x}}^{(c)}$. Furthermore, $\Pr(\mathrm{CPL}|\mathcal{C}_p)$ is a function that depends on the correct path $\mathbf{x}^{(c)}$ and, strictly speaking, has to be averaged over all correct paths. We shall be satisfied with the correct path that produces the largest Pr(CPL).

In Appendix 6.A we show that the probability of losing the correct path decreases exponentially with the signal-to-noise ratio and is bounded above by

$$\Pr(\text{CPL}|\mathcal{C}_p) \le Q\left(\sqrt{\frac{d_l^2}{2N_0}}\right). \tag{6.21}$$

The parameter d_l^2 depends on $\mathcal{C}_p$ and is known as the *vector Euclidean distance* [15] of the path $\tilde{\mathbf{x}}^{(c)}$ with respect to the M error paths $\tilde{\mathbf{x}}^{(p_i)} \in \mathcal{C}_p$. Note that (6.21) is an upper bound on the probability that M specific error paths have a metric larger than $\tilde{\mathbf{x}}^{(c)}$. Finding d_l^2 involves a combinatorial search (see Appendix 6.A).

Equation (6.21) demonstrates that the probability of correct path loss is an exponential error integral and can thus be compared to the probability of the ML decoder (equations (5.8) and (5.9)). The problem is finding the minimal d_l^2 among all sets $\mathcal{C}_p$, denoted by $\min(d_l^2)$. This is a rather complicated combinatorial problem, since essentially all combinations of M candidates for each correct path at each time n have to be analyzed from the growing set of possibilities. Aulin [15] has studied this problem and gives several rejection rules that alleviate the complexity of finding d_l^2, but the problem remains very complex and is in need of further study.

Note that d_l^2 is a nondecreasing function of M, the decoder complexity, and one way of selecting M is to choose it such that

$$\min(d_l^2) \ge d_{\text{free}}^2. \tag{6.22}$$

This choice should guarantee that the performance of the M-algorithm is approximately equal to the performance of ML decoding. To see this, let $\overline{P_e(M)}$ be the probability of an error event (compare equation (5.4)). Then

$$\overline{P_e(M)} \le \overline{P_e}\,(1 - \Pr(\text{CPL})) + \Pr(\text{CPL})$$
$$\le \overline{P_e} + \Pr(\text{CPL}), \tag{6.23}$$

where $\overline{P_e}$ is of course the probability that an ML decoder starts an error event (Chapter 5). For high values of the signal-to-noise ratio, equation (6.23) can be approximated by

$$\overline{P_e(M)} \approx N_{d_{\text{free}}} Q\left(\frac{d_{\text{free}}}{\sqrt{2N_0}}\right) + \kappa Q\left(\frac{\min(d_l)}{\sqrt{2N_0}}\right), \tag{6.24}$$

where κ is some constant, which, however, is difficult to determine in the general case. Now, if (6.22) holds, the suboptimality does not exponentially dominate the error performance for high signal-to-noise ratios.

Aulin [15] has analyzed this situation for 8-PSK trellis codes and found that, in general, $M \approx \sqrt{S}$ will fulfill condition (6.22), where S is the number of code states. Since the complexity of the M-algorithm is independent of the code size, one would want to use codes that are large and powerful and have a large free-squared Euclidean distance d_{free}^2. However, since the probability of path loss and the probability of an ML decoding error can be made asymptotically equivalent, it makes no sense to use a code that is too powerful. Doing so only presents the algorithm with more resynchronization difficulties.

Figure 6.5 shows the simulated performance of the M-algorithm versus M for the 64-state trellis code from Table 3.1 with $d_{\text{free}}^2 = 6.34$. $M = 8$ meets (6.22) according to [15], but from Figure 6.5 it is apparent that the performance is still about 1.5 dB poorer than ML decoding. This is attributable to the resynchronization problems and the fact that we are operating at rather low values of the signal-to-noise ratio, where neither d_{free}^2 nor $\min(d_l^2)$ are necessarily dominating the error performance.

Figures 6.6 and 6.7 show the empirical probability of correct path loss Pr(CPL) and the bit error probability (BER) for two convolutional

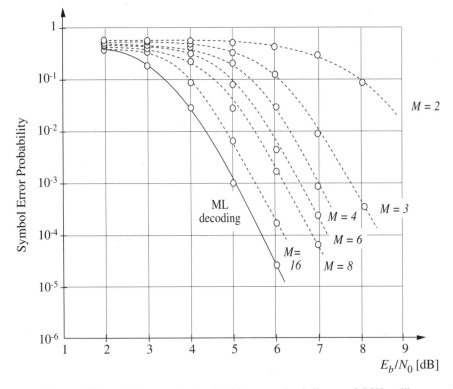

Figure 6.5 Simulation results for the 64-state optimal distance 8-PSK trellis codes decoded with the M-algorithm, using $M = 2, 3, 4, 5, 6, 8,$ and 16 (from [15]). The performance of maximum-likelihood decoding is also included in the figure.

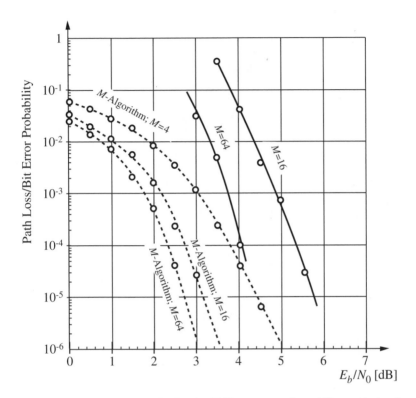

Figure 6.6 Simulation results for the 2048-state, rate $R = 1/2$ convolutional
code using the M-algorithm, for $M = 4, 16$, and 64. The dashed
curves are the path loss probability, and the solid curves are BERs.

codes and various values of M. Figure 6.6 shows simulation results for
the 2048-state convolutional code, $v = 11$, from Table 4.1. The bit error
rate and the probability of losing the correct path converge to the same
asymptotic behavior, indicating that the probability of correct path loss
and not recovery errors is the dominant error mechanism for very large
values of the signal-to-noise ratio.

Figure 6.7 shows the same simulation results for the $v = 15$ large-
constraint-length code for the same values of M. For this length code, path
loss will be the dominant error scenario. We note that both codes have
a very similar error performance, demonstrating that the code complexity
has little influence.

Once the correct path is lost, the algorithm may spend a relatively
long time before it finds it again, that is, before the correct path is again
one of the M retained paths. Correct path recovery is a very complex
problem, and no complete analytical results have been found. There are

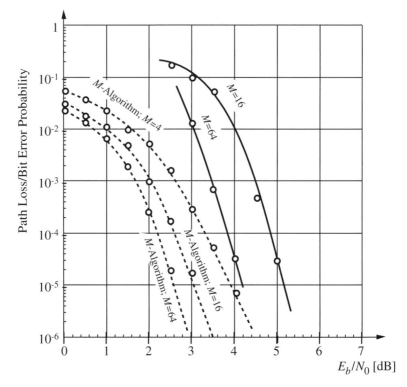

Figure 6.7 Same simulation results for the $\nu = 15$, $R = 1/2$ convolutional code.

only a few theoretical approaches to the recovery problem, such as [16]. This difficulty suggests that some insight into the operation of the decoder during a recovery may be gained through simulation studies.

Figure 6.8 shows the simulated average number of steps taken for the algorithm to recover the correct path. The simulations were done for the 2048-state, $\nu = 11$ code, whose error performance is shown in Figure 6.6. Each instance of the simulation was performed such that the algorithm was initiated and run until the correct path was lost. Then the number of steps until recovery were counted [17].

Figure 6.9 shows the average number of steps until recovery for the rate-$1/2$, $\nu = 11$ systematic convolutional code with generator polynomials $g^{(0)} = 4000$, $g^{(1)} = 7153$. This code has a free Hamming distance of only $d_{\text{free}} = 9$, but its recovery performance is much superior to that of the nonsystematic code. In fact, the average number of steps until recovery is independent of the signal-to-noise ratio, but it increases approximately linearly with E_b/N_0 for the nonsystematic code. This rapid recovery results in superior error performance of the systematic code compared to the

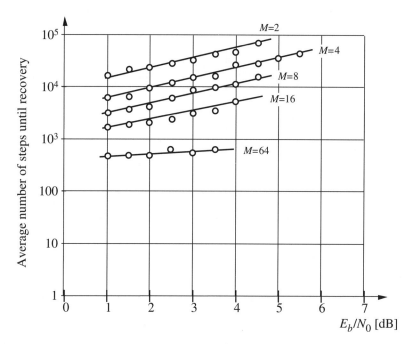

Figure 6.8 Average number of steps until recovery of the correct path for the
code from Figure 6.6 [17].

nonsystematic code, shown in Figure 6.10, even though its free distance
is significantly smaller. What is true, however, and can be seen clearly in
Figure 6.10, is that for very large values of E_b/N_0 the "stronger" code will
win out due to its larger free distance.

The observation that systematic convolutional codes outperform nonsystematic codes for error rates $P_b \gtrsim 10^{-6}$ has also been made by Osthoff et al. [18]. The reason for this difference lies in the *return barrier* phenomenon, which can be explained with the aid of Figure 6.11.
In order for the algorithm to recapture the correct path after a correct
path loss, one of the M retained paths must correspond to a trellis state
with a connection to the correct state at the next time interval. In Figure
6.11 we assume that the all-zero sequence is the correct sequence, and
hence the all-zero state is the correct state for all time intervals. This assumption is made without loss of generality for convolutional codes due
to their linearity. For a feedforward realization of a rate-1/2 code, the
only state that connects to the all-zero state is $(0, \ldots, 0, 1)$, denoted by s_m
in the figure. For a systematic code with $g_0^{(1)} = g_v^{(1)} = 1$ (see Chapter
4) the two possible branch signals are (01) and (10), as indicated in
Figure 6.11.

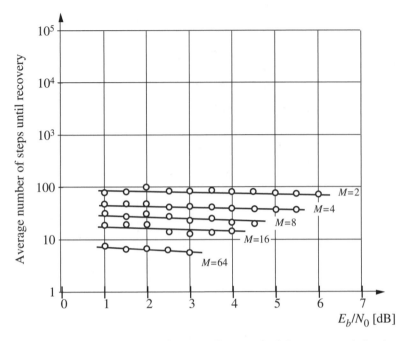

Figure 6.9 Average number of steps until recovery of the correct path for the systematic convolutional code with constraint length $\nu = 11$ [17].

For a nonsystematic, maximum free-distance code, on the other hand, the two branch signals are (11) and (00), respectively. Since the correct branch signal is (00), the probability that the metric of s_f (for failed) exceeds the metric of s_c is $1/2$ for the systematic code, since both branch signals are equidistant from the correct branch signal. For the nonsystematic code, on the other hand, this probability is $Q(\sqrt{E_b/2N_0})$. This explains the dependence of the path recovery on E_b/N_0 for nonsystematic codes, as well as why systematic codes recapture the correct path faster with a recovery behavior independent of E_b/N_0.

The M-algorithm impresses with its simplicity. Unfortunately, a theoretical understanding of the algorithm is not related to this simplicity at all, and it seems that much more work in this area is needed before a coherent theory is available. This lack of a theoretical basis for the algorithm is, however, no barrier to its implementation. Early work on the application of the M-algorithm to convolutional codes was presented by Zigangirov and Kolesnik [19], and Simmons and Wittke [20], Aulin [21], and Balachandran [22], among others, have applied the M-algorithm to continuous-phase modulation. General trellis codes have not yet seen much action from the M-algorithm. A notable exception is [9].

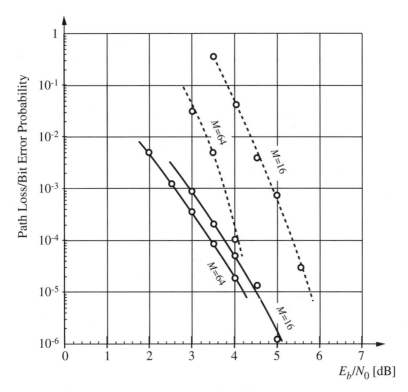

Figure 6.10 Simulation results for the superior 2048-state systematic code using the M-algorithm. The dashed curves are the error performance of the same-constraint-length nonsystematic code from Figure 6.6 [17].

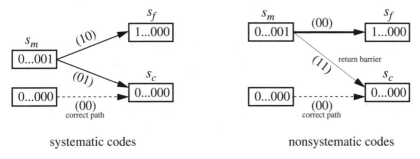

Figure 6.11 Heuristic explanation of the return barrier phenomenon in the M-algorithm.

6.5   MAXIMUM-LIKELIHOOD DECODING

The difficulty in decoding trellis codes arises from the exponential size of the growing decoding tree. In this section we show that this tree can be reduced by merging nodes, such that the tree only grows to a maximum size of 2^S nodes, where S is the number of encoder states. This merging leads diverging paths together again, and we obtain a structure resembling a trellis, as discussed for encoders in Section 3.

To see how this happens, let $J_{n-1}^{(i)}$ and $J_{n-1}^{(j)}$ be the metrics of two nodes corresponding to the partial sequences $\tilde{\mathbf{x}}^{(i)}$ and $\tilde{\mathbf{x}}^{(j)}$ of length $n-1$, respectively. Let the encoder states that correspond to $\tilde{\mathbf{x}}^{(i)}$ and $\tilde{\mathbf{x}}^{(j)}$ after time $n-1$ be $s_n^{(i)}$ and $s_n^{(j)}$, $s_n^{(i)} \neq s_n^{(j)}$, and assume that the next extension of $\tilde{\mathbf{x}}^{(i)} \to (\tilde{\mathbf{x}}^{(i)}, x_n^{(i)})$ and $\tilde{\mathbf{x}}^{(j)} \to (\tilde{\mathbf{x}}^{(j)}, x_n^{(j)})$ is such that $s_{n+1}^{(i)} = s_{n+1}^{(j)}$; that is, the encoder states at time n are identical. See also Figure 6.12.

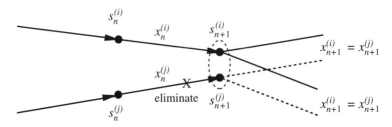

Figure 6.12 Merging nodes.

Now we propose to merge nodes $(\tilde{\mathbf{x}}^{(i)}, x_n^{(i)})$ and $(\tilde{\mathbf{x}}^{(j)}, x_n^{(j)})$ into one node, which we now call a *state*. We retain the partial sequence with the larger metric J_n at time n and discard the partial sequence with the smaller metric. Ties are broken arbitrarily. We are now ready to prove the following

THEOREM 6.1
THEOREM OF NONOPTIMALITY: The procedure of merging nodes that correspond to identical encoder states and discarding the path with the smaller metric never eliminates the maximum-likelihood path.

Theorem 6.1 is sometimes referred to as the theorem of nonoptimality and allows us to construct an ML decoder whose complexity is significantly smaller than that of an all-out exhaustive tree search.

PROOF
The metric at time $n + k$ can be written as

$$J_{n+k}^{(i)} = J_n^{(i)} + \sum_{h=1}^{k} \beta_{n+h}^{(i)} \tag{6.25}$$

for every future metric $J_{n+k}^{(i)}$, $0 < k \leq l - n$, where $\beta_{n+h}^{(i)} = 2\text{Re}\left\{x_n^{(i)} y_n^*\right\} - |x_n^{(i)}|^2$ is the metric increment, now also called the *branch metric*, at time n.

Now, if the nodes of paths i and j correspond to the same encoder state at time n, there exists for every possible extension of the ith path $(x_n^{(i)}, \ldots, x_{n+k}^{(i)})$ a corresponding identical extension $(x_n^{(j)}, \ldots, x_{n+k}^{(j)})$ of the jth path, and vice versa. Let us then assume without loss of generality that the ith path accumulates the largest metric at time l; i.e., $J_l^{(i)} \geq J_l^{(j)}$. Therefore,

$$J_n^{(i)} + \sum_{h=1}^{l} \beta_{n+h}^{(i)} \geq J_n^{(j)} + \sum_{h=1}^{l} \beta_{n+h}^{(j)}, \qquad (6.26)$$

and $\sum_{h=1}^{l} \beta_{n+h}^{(i)}$ is the maximum metric sum for the extensions of the ith path. (Otherwise another path would have a higher final metric.) But since the extensions from $s_n^{(i)}$ and $s_n^{(j)}$ are identical, $\sum_{h=1}^{l} \beta_{n+h}^{(i)} = \sum_{h=1}^{l} \beta_{n+h}^{(j)}$ and $J_n^{(i)} \geq J_n^{(j)}$. Path j can therefore never accumulate a larger metric than path i and we may discard it with impunity at time n.

The tree now folds back on itself and forms a trellis with exactly S states[3] (see also Figure 3.3), and there are 2^k paths merging in a single state at each step. Note then that there are now at most S retained partial sequences $\tilde{\mathbf{x}}^{(i)}$, called the *survivors*. The most convenient labeling convention is that each state is labeled by the corresponding encoder state plus the survivor that leads to it. This trellis is an exact replica of the encoder trellis discussed in Chapter 3, and the task of the decoder is to retrace the path the encoder traced through this trellis. Theorem 6.1 guarantees that this procedure is optimal. This method was introduced by Viterbi in 1967 [23, 24] in the context of analyzing convolutional codes and has since become widely known as the *Viterbi algorithm* [25]:

Step 1.   Initialize the S states of the ML decoder with a metric $J_{-l}^{(i)} = -\infty$ and survivors $\tilde{\mathbf{x}}^{(i)} = \{\ \}$. Initialize the starting state of the encoder, usually state $i = 0$, with the metric $J_{-l}^{(0)} = 0$. Let $n = -l$.

Step 2.   Calculate the branch metric

$$\beta_n = 2\mathrm{Re}\left\{x_n y_n^*\right\} - |x_n|^2 \qquad (6.27)$$

for each state $s_n^{(i)}$ and each extension $x_n^{(i)}$.

Step 3.   Shift the states $s_n^{(i)} \rightarrow s_{n+1}^{(i)}$ according to the trellis transitions determined by the encoder FSM and, from the 2^k merging paths, retain the survivor $\tilde{\mathbf{x}}^{(i)}$ for which $J_n^{(i)}$ is maximized.

Step 4.   If $n < l$, let $n = n + 1$ and go to Step 2.

Step 5.   Output the survivor $\mathbf{x}^{(i)}$ that maximizes $J_l^{(i)}$ as the ML estimate of the transmitted sequence.

---

[3] In the context of the trellis we refer to the merged nodes as states.

The Viterbi algorithm and the M-algorithm are both breadth-first searches and share some similarities. In fact, one often introduces the concept of mergers also in the M-algorithm to avoid carrying along suboptimal paths. In fact, the M-algorithm can be operated in the trellis rather than in the tree. The Viterbi algorithm has enjoyed tremendous popularity not only in decoding trellis codes but also in symbol sequence estimation over channels affected by intersymbol interference [26, 27] and multiuser optimal detectors [28]. Whenever the underlying generating process can be modeled as a finite-state machine, the Viterbi algorithm finds application.

A rather large body of literature deals with the Viterbi decoder, and some good books deal with the subject (e.g., [6, 26, 29, 30]). One of the more important results is that it can be shown that one does not have to wait until the entire sequence is decoded before starting to output the estimated symbols $x_n^{(i)}$ or the corresponding data. The probability that the symbols in all survivors $\tilde{\mathbf{x}}^{(i)}$ are identical for $m < n - n_t$, where n is the current active decoding time and n_t, called the *truncation length* or *decision depth*, is $\approx 5v$, the constraint length of the code (equation (4.16)) is very close to unity for $n_t \approx 5v$. This has been shown to be true for rate-1/2 convolutional codes [31, page 182], but the argument can easily be extended to general trellis codes. We may therefore modify the algorithm to obtain a fixed-delay decoder by modifying Steps 4 and 5 of the Viterbi algorithm as follows:

Step 4. If $n \geq n_t$, output $x_{n-n_t}^{(i)}$ from the survivor $\tilde{\mathbf{x}}^{(i)}$ with the largest metric $J_n^{(i)}$ as the estimated symbol at time $n - n_t$. If $n < l - 1$, let $n = n + 1$ and go to Step 2.

Step 5. Output the remaining estimated symbols $x_n^{(i)}$, $l - n_t < n \leq l$, from the survivor $\mathbf{x}^{(i)}$ that maximizes $J_l^{(i)}$.

We recognize that we may now let $l \to \infty$; that is, the complexity of our decoder is no longer determined by the length of the sequence, and it may be operated in a continuous fashion. The simulation results in Chapter 5 were obtained with a Viterbi decoder according to the modified algorithm.

Let us think about the complexity of the Viterbi algorithm. Denote by E the total number of branches in the trellis on which we wish to operate the Viterbi algorithm; i.e., for a linear trellis there are $S2^k$ branches per time epoch. The complexity requirements of the Viterbi algorithm can then be captured by the following [32].

THEOREM 6.2

The Viterbi algorithm requires a complexity that is linear in the number of edges E; that is, it performs $O(E)$ arithmetic operations (multiplications, additions, and comparisons).

PROOF

Step 2 in the Viterbi algorithm requires the calculation of β_n, which needs two multiplies and an addition, and the addition $J_n^{(i)} + \beta_n$ for each branch. Some of the values β_n may be identical, the number of arithmetic operations is therefore larger than E additions and less than $2E$ multiplications and additions.

   If we denote the number of branches entering state s by $\rho(s)$, Step 3 requires $\sum_{\text{states } s}(\rho(s) - 1) \leq E/2l$ comparisons per time epoch. $\rho(s) = 2^k$ in our case, and the total number of comparisons is therefore less than E and larger than $E - 2lS$. Then $O(E)$ arithmetic operations are required.

6.6   MAXIMUM A POSTERIORI SYMBOL DECODING

Maximum-likelihood decoding minimizes the probability of an error event. Although this also ensures a small symbol or bit error probability, ML decoding is not the exact solution to the problem of minimizing the bit error probability. This is the task of the maximum a posteriori (MAP) decoder discussed in this section. The ML decoder is not optimal in the sense of bit error probability; however, in most applications, its performance is virtually identical to that of the MAP decoder. As we will see, the complexity of the MAP decoder is significantly higher than that of the ML decoder, and it is therefore not an attractive alternative. Its main virtue is that it produces, as a by-product, the a posteriori channel symbol and information bit probabilities. These probabilities are required for iterative decoding schemes and concatenated coding schemes with soft-decision decoding of the inner code, most notably for iterative decoding of Turbo codes, which is discussed in Chapter 8. The derivation of the MAP decoder presented in this section was first given by Bahl et al. [33].

   Recall that the encoder of a trellis code is a finite-state machine (FSM) driven by the input information bits. The transitions of the FSM at time r are governed by the transition probabilities

$$p_{ij} = \Pr(s_{r+1} = j | s_r = i) = \Pr(u_r), \tag{6.28}$$

where u_r causes the transition from state i to state j. Furthermore,

$$q_{ij}(x) = \Pr(\tau(u_r, s_r) = x | s_r = i, s_{r+1} = j) \tag{6.29}$$

is the a priori probability that the output x_r at time r assumes the value x on the transition from state i to state j. This probability is a deterministic function of i and j unless there are parallel transitions, in which case x_r is determined by the uncoded information bits (see Section 3.4). The encoded sequence from the encoder, $\mathbf{x}$, passes through the transmission channel at whose output we observe the received sequence $\mathbf{y}$. A first task of the MAP decoder is to estimate the a posteriori probabilities of the encoder states and the transitions of the FSM from the complete received sequence $\mathbf{y}$—calculate

$$\Pr(s_{r+1} = j | \mathbf{y}), \tag{6.30}$$

and the transition probabilities

$$\Pr(s_r = i; s_{r+1} = j | \mathbf{y}) \tag{6.31}$$

for every r.

It turns out to be easier to evaluate the joint probabilities

$$\Pr(s_{r+1} = j, \mathbf{y}) = \Pr(s_{r+1} = j | \mathbf{y})\Pr(\mathbf{y}), \tag{6.32}$$

and

$$\Pr(s_r = i, s_{r+1} = j, \mathbf{y}) = \Pr(s_r = i, s_{r+1} = j, \mathbf{y})\Pr(\mathbf{y}), \tag{6.33}$$

from which (6.30) and (6.31) can be reconstructed easily by dividing by $\Pr(\mathbf{y})$, which can be accomplished simply by normalizing (6.32) and (6.33) to sum to unity.

Before we proceed with the derivation, we define some additional auxiliary probability functions. These are

$$\alpha_r(j) = \Pr(s_{r+1} = j, \tilde{\mathbf{y}}), \tag{6.34}$$

the joint probability of the partial sequence $\tilde{\mathbf{y}} = (y_{-l}, \ldots, y_r)$ up to and including time epoch r and state $s_{r+1} = j$;

$$\beta_r(j) = \Pr((y_{r+1}, \ldots, y_l) | s_{r+1} = j), \tag{6.35}$$

the conditional probability of the remainder of the received sequence $\mathbf{y}$ given that the state at time $r + 1$ is j; and

$$\gamma_r(j, i) = \Pr(s_{r+1} = j, y_r | s_r = i), \tag{6.36}$$

the joint conditional probability of y_r and that the state at time $r + 1$ equals j, given that the state at time r is i.

With the above we now calculate

$$\text{Pr}(s_{r+1} = j, \mathbf{y}) = \text{Pr}(s_{r+1} = j, \tilde{\mathbf{y}}, (y_{r+1}, \ldots, y_l))$$
$$= \text{Pr}(s_{r+1} = j, \tilde{\mathbf{y}})\text{Pr}((y_{r+1}, \ldots, y_l)|s_{r+1} = j, \tilde{\mathbf{y}}) \quad (6.37)$$
$$= \alpha_r(j)\beta_r(j),$$

where we have used the fact that $\text{Pr}((y_{r+1}, \ldots, y_l)|s_{r+1} = j, \tilde{\mathbf{y}}) = \text{Pr}((y_{r+1}, \ldots, y_l)|s_{r+1} = j)$; i.e., if $s_{r+1} = j$ is known, events after time r are independent of the history $\tilde{\mathbf{y}}$ up to s_{r+1}.

In the same way we calculate via Bayes' expansion

$$\text{Pr}(s_r = i, s_{r+1} = j, \mathbf{y}) = \text{Pr}(s_r = i, s_{r+1} = j, (y_{-l}, \ldots, y_{r-1}), y_r, (y_{r+1}, \ldots, y_l))$$
$$= \text{Pr}(s_r = i, (y_{-l}, \ldots, y_{r-1}))\text{Pr}(s_{r+1} = j, y_r|s_r = i)$$
$$\times \text{Pr}((y_{r+1}, \ldots, y_l)|s_{r+1} = j)$$
$$= \alpha_{r-1}(i)\gamma_r(j, i)\beta_r(j).$$
$$(6.38)$$

Now, again applying Bayes' rule and $\sum_b p(a, b) = p(a)$, we obtain

$$\alpha_r(j) = \sum_{\text{states } i} \text{Pr}(s_r = i, s_{r+1} = j, \tilde{\mathbf{y}})$$
$$= \sum_{\text{states } i} \text{Pr}(s_r = i, (y_{-l}, \ldots, y_{r-1}))\text{Pr}(s_{r+1} = j, y_r|s_r = i)$$
$$= \sum_{\text{states } i} \alpha_{r-1}(i)\gamma_r(j, i). \quad (6.39)$$

For a trellis code started in the zero state at time $r = -l$ we have the starting conditions

$$\alpha_{-l-1}(0) = 1, \alpha_{-l-1}(j) = 0, \qquad j \neq 0. \quad (6.40)$$

As above, we similarly develop an expression for $\beta_r(j)$:

$$\beta_r(j) = \sum_{\text{states } i} \text{Pr}(s_{r+2} = i, (y_{r+1}, \ldots, y_l)|s_{r+1} = j)$$
$$= \sum_{\text{states } i} \text{Pr}(s_{r+2} = i, y_{r+1}|s_{r+1} = j)\text{Pr}((y_{r+2}, \ldots, y_l)|s_{r+2} = i) \quad (6.41)$$
$$= \sum_{\text{states } i} \beta_{r+1}(i)\gamma_{r+1}(i, j).$$

The boundary condition for $\beta_r(j)$ is

$$\beta_l(0) = 1, \beta_l(j) = 0, \qquad j \neq 0, \quad (6.42)$$

for a trellis code terminated in the zero state.

Furthermore,

$$\gamma_r(j, i) = \sum_{x_r} \Pr(s_{r+1} = j | s_r = i) \Pr(x_r | s_r = i, s_{r+1} = j) \Pr(y_r | x_r)$$

$$= \sum_{x_r} p_{ij} q_{ij}(x_r) p_n(y_r - x_r), \tag{6.43}$$

where we have used the conditional density function of the additive white Gaussian noise (AWGN) channel from (2.11), namely $\Pr(y_r | x_r) = p_n(y_r - x_r)$. The calculation of $\gamma_r(j, i)$ is not very complex and can most easily be implemented by a table look-up procedure.

Equations (6.39) and (6.41) are iterative, and we can now compute the a posteriori state and transition probabilities via the following algorithm:

Step 1. Initialize $\alpha_{-l-1}(0) = 1$, $\alpha_{-l-1}(j) = 0$ for all nonzero states ($j \neq 0$) of the encoder FSM, and $\beta_l(0) = 1$, $\beta_l(j) = 0$, $j \neq 0$. Let $r = -l$.

Step 2. Calculate $\gamma_r(j, i)$ using (6.43) and $\alpha_r(j)$ using (6.39) for all states j.

Step 3. If $r < l$, let $r = r + 1$ and go to Step 2; else $r = l - 1$ and go to Step 4.

Step 4. Calculate $\beta_r(j)$ using (6.41). Calculate $\Pr(s_{r+1} = j, \mathbf{y})$ from (6.37) and $\Pr(s_r = i, s_{r+1} = j; \mathbf{y})$ from (6.38).

Step 5. If $r > -l$, let $r = r - 1$ and go to Step 4.

Step 6. Terminate the algorithm and output all the values $\Pr(s_{r+1} = j, \mathbf{y})$ and $\Pr(s_r = i, s_{r+1} = j, \mathbf{y})$.

Contrary to the ML algorithm, the MAP algorithms needs to go through the trellis twice, once in the forward direction and once in the reverse direction. What is worse, all the values $\alpha_r(j)$ must be stored from the first pass through the trellis. For a rate-k/n convolutional code, for example, this requires $2^{k\nu} 2l$ storage locations since there are $2^{k\nu}$ states, for each of which we need to store a different value $\alpha_r(j)$ at each time epoch r. The storage requirement grows exponentially in the constraint length ν and linearly in the block length $2l$.

The a posteriori state and transition probabilities produced by this algorithm can now be used to calculate a posteriori information bit probabilities—i.e., the probability that the information k-tuple $u_r = u$, where u can vary over all possible binary k-tuples. Starting from the transition probabilities $\Pr(s_r = i, s_{r+1} = j | \mathbf{y})$, we simply sum over all transitions $i \rightarrow j$ caused by $u_r = u$. Denoting these transitions by $A(u)$, we obtain

$$\Pr(u_r = u) = \sum_{(i,j) \in A(u)} \Pr(s_r = i, s_{r+1} = j | \mathbf{y}). \qquad (6.44)$$

As mentioned, another most interesting product of the MAP decoder is the a posteriori probability of the transmitted output symbol x_r. Arguing analogously as before and letting $B(x)$ be the set of transitions on which the output signal x can occur, we obtain

$$\Pr(x_r = x) = \sum_{(i,j) \in B(x)} \Pr(x | y_r) \Pr(s_r = i, s_{r+1} = j | \mathbf{y})$$

$$= \sum_{(i,j) \in B(x)} \frac{p_n(y_r - x_r)}{p(y_r)} q_{ij}(x) \Pr(s_r = i, s_{r+1} = j | \mathbf{y}), \quad (6.45)$$

where the a priori probability of v_r can be calculated via

$$p(y_r) = \sum_{x'} p(y_r | x') q_{ij}(x'). \qquad (6.46)$$

Equation (6.45) can be much simplified if there is only one output symbol on the transition $i \rightarrow j$. In this case the transition automatically determines the output symbol, and

$$\Pr(x_r = x) = \sum_{(i,j) \in B(x)} \Pr(s_r = i, s_{r+1} = j | \mathbf{v}). \qquad (6.47)$$

The algorithm in this section is not a viable alternative to ML decoding due to its larger complexity. However, the decoding of Turbo codes in Chapter 8 requires the a posteriori symbol probabilities $\Pr(x_r = x)$, which the MAP decoder can furnish, and we use such a MAP decoder as one part of an iterative decoder.

6.7   RANDOM CODING ANALYSIS OF SEQUENTIAL DECODING

In Section 5.5 we presented random coding performance bounds for trellis codes. We implicitly assumed that we were using an ML decoder, such as the one discussed in the preceding section. Since sequential decoding is not an ML decoding method, the results in Section 5.5 may not apply.

The error analysis of sequential decoding is very difficult, and, again, we find it easier to generate results for the ensemble of all trellis codes via random coding arguments. The evaluation of the error probability is not the main problem here, since, if properly dimensioned, both the stack and the Fano algorithm will almost always find the same error path as the ML detector.

The difference with sequential decoding is, in contrast to ML decoding, that its computational load is variable. And it is this computational load that can cause problems, as we will see. Figure 6.13 shows an example of the search procedure of sequential decoding. The algorithm explores at each node an entire set of incorrect partial paths before finally continuing. This set at each node includes all the incorrect paths explored by the possibly multiple visits to that node as, for example, in the Fano algorithm. The sets $\mathcal{X}'_j$ denote the sets of incorrect signal sequences $\tilde{\mathbf{x}}'$ diverging at node j that are searched by the algorithm. Further, denote the number of signal sequences in $\mathcal{X}'_j$ by C_j. Note that C_j is also the number of computations that need to be done at node j, since each new path requires one additional metric calculation.

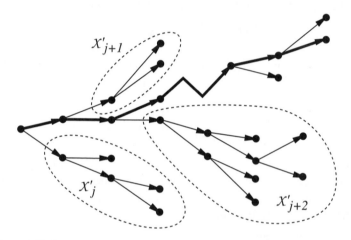

Figure 6.13 Incorrect subsets explored by a sequential decoder. The solid path is the correct one.

The problem becomes quite evident now. The number of computations at each node is variable, and it is this distribution of the computations that we want to examine. Again, let $\tilde{\mathbf{x}}$ be the partial correct path through the trellis and $\tilde{\mathbf{x}}'_j$ a partial incorrect path that diverges from $\tilde{\mathbf{x}}$ at node j. Furthermore, let $L_n(\tilde{\mathbf{x}}') = L(\tilde{\mathbf{x}}', \mathbf{y})$ be the metric of the incorrect path at node n, and let $L_m(\tilde{\mathbf{x}})$ be the metric of the correct path at node m. A path is searched further if and only if it is at the top of the stack; hence, if $L_n(\tilde{\mathbf{x}}') < L_m(\tilde{\mathbf{x}})$, the incorrect path is not searched further until the metric of $\tilde{\mathbf{x}}$ falls below $L_n(\tilde{\mathbf{x}}')$. If

$$\min_{m \geq j} L_m(\tilde{\mathbf{x}}) = \lambda_j > L_n(\tilde{\mathbf{x}}') \tag{6.48}$$

the incorrect path $\tilde{\mathbf{x}}'$ is never searched beyond node n.

We may now give an upper bound to the probability that the number of computations at node j exceeds a given value N_c by

$$\Pr(C_j \geq N_c) \leq \sum_{\mathbf{x}} p(\mathbf{x}) \int_{\mathbf{y}} p(\mathbf{y}|\mathbf{x}) \mathcal{B}\left(|e(p(\mathbf{y}|\mathbf{x}') \geq \lambda_j)| \geq N_c\right) \, d\mathbf{y}, \quad (6.49)$$

where $e(p(\mathbf{y}|\mathbf{x}') \geq \lambda_j)$ is an error path in $\mathcal{X}_j$ whose metric exceeds λ_j and $|\ |$ is the number of such error paths. $\mathcal{B}(\star)$ is a Boolean function that equals 1 if the expression is true, and is 0 otherwise. The function $\mathcal{B}(\star)$ in (6.49) then simply equals 1 if there are more than N_c error paths with metric larger than λ_i and 0 otherwise.

We now proceed to bound above the indicator function analogously to Chapter 5 by realizing that $\mathcal{B}(\star) = 1$ if at least N_c error paths have a metric such that

$$L_n(\tilde{\mathbf{x}}') \geq \lambda_j, \quad (6.50)$$

and, hence,

$$\left(\frac{1}{N_c} \sum_{\tilde{\mathbf{x}}' \in \mathcal{X}_j'} \exp\left(\alpha \left(L_n(\tilde{\mathbf{x}}') - \lambda_j\right)\right)\right)^{\rho} \geq 1, \quad (6.51)$$

where α and ρ are arbitrary positive constants. Note that we have extended the sum in (6.51) over all error sequences, as is customary in random coding analysis. We may now use (6.51) to bound the indicator function $\mathcal{B}(\star)$ above, and we obtain

$$\Pr(C_j \geq N_c) \leq N_c^{-\rho} \sum_{\mathbf{x}} p(\mathbf{x}) \int_{\mathbf{y}} p(\mathbf{y}|\mathbf{x}) \left(\sum_{\tilde{\mathbf{x}}' \in \mathcal{X}_j'} \exp\left(\alpha(L_n(\tilde{\mathbf{x}}') - \lambda_j)\right)\right)^{\rho} d\mathbf{y}, \quad (6.52)$$

and, due to (6.48),

$$\exp\left(-\alpha \rho \lambda_i\right) \leq \sum_{m=j}^{\infty} \exp\left(-\alpha \rho L_m(\tilde{\mathbf{x}})\right), \quad (6.53)$$

and we have

$$\Pr(C_j \geq N_c) \leq N_c^{-\rho} \sum_{\mathbf{x}} p(\mathbf{x}) \int_{\mathbf{y}} p(\mathbf{y}|\mathbf{x}) \left(\sum_{\tilde{\mathbf{x}}' \in \mathcal{X}_j'} \exp\left(\alpha L_n(\tilde{\mathbf{x}}')\right)\right)^{\rho}$$
$$\times \sum_{m=j}^{\infty} \exp\left(-\alpha \rho L_m(\tilde{\mathbf{x}})\right) \, d\mathbf{y}. \quad (6.54)$$

Analogously to Chapter 5, let c be the correct path and e the incorrect path that diverges from c at node j. Let $\mathcal{E}$ be the set of all incorrect paths, and $\mathcal{E}'_j$ the set of incorrect paths (not signal sequences) corresponding to $\mathbf{x}'_j$. Again, $\mathbf{x}$ and $\mathbf{x}'$ are, strictly taken, the signal sequences assigned to the correct and incorrect path, respectively, and $\tilde{\mathbf{x}}, \tilde{\mathbf{x}}'$ are the associated partial sequences. Let $\mathrm{Avg}\{\mathrm{Pr}(C_j \geq N_c)\}$ be the ensemble average of $\mathrm{Pr}(C_j \geq N_c)$ over all linear trellis codes:

$$\mathrm{Avg}\{\mathrm{Pr}(C_j \geq N_c)\} \leq N_c^{-\rho} \sum_c p(c) \int_{\mathbf{y}} p(\mathbf{y}|\mathbf{x}) \sum_{m=j}^{\infty} \exp\left(-\alpha\rho L_m(\tilde{\mathbf{x}})\right)$$

$$\times \left(\sum_{e \in \mathcal{E}'_j} \overline{\exp\left(\alpha L_n(\tilde{\mathbf{x}}')\right)}\right)^{\rho} \, d\mathbf{y}. \qquad (6.55)$$

We have used Jensen's inequality to pull the averaging into the second sum, which restricts ρ to $0 \leq \rho \leq 1$. Since we are using time-varying random trellis codes, (6.55) becomes independent of the starting node j, which we arbitrarily set to $j = 0$ now.

Observe there are at most 2^{kn} paths e of length n in $\mathcal{E}'_j = \mathcal{E}'$. Using this and the inequality[4]

$$\left(\sum a_i\right)^{\rho} \leq \sum a_i^{\rho}, \qquad a_i \geq 0, \quad 0 \leq \rho \leq 1, \qquad (6.56)$$

we obtain[5]

$$\mathrm{Avg}\{\mathrm{Pr}(C_0 \geq N_c)\} \leq N_c^{-\rho} \sum_c p(c) \int_{\mathbf{y}} p(\mathbf{y}|\mathbf{x}) \sum_{m=0}^{\infty} \exp\left(-\alpha\rho L_m(\tilde{\mathbf{x}})\right)$$

$$\times \sum_{n=0}^{\infty} 2^{kn\rho} \left(\overline{\exp\left(\alpha L_n(\tilde{\mathbf{x}}')\right)}\right)^{\rho} \, d\mathbf{y} \qquad (6.57)$$

$$= N_c^{-\rho} \sum_c p(c) \sum_{m=0}^{\infty} \sum_{n=0}^{\infty} 2^{kn\rho}$$

$$\times \overline{\int_{\mathbf{y}} p(\mathbf{y}|\mathbf{x}) \exp\left(-\alpha\rho L_m(\tilde{\mathbf{x}})\right) \left(\overline{\exp\left(\alpha L_n(\tilde{\mathbf{x}}')\right)}\right)^{\rho}} \, d\mathbf{y}. (6.58)$$

---

[4] This inequality is easily shown, that is,

$$\frac{\sum a_i^{\rho}}{\left(\sum a_i\right)^{\rho}} = \sum \left(\frac{a_i}{\sum a_i}\right)^{\rho} \geq \sum \left(\frac{a_i}{\sum a_i}\right) = 1,$$

where the inequality resulted from the fact that each term in the sum is ≤ 1 and $\rho \leq 1$.

[5] Note that it is here that we need the time-varying assumption of the codes (compare also Section 5.5 and Figure 5.9).

Now we substitute the metrics (see (6.11))

$$L_m(\tilde{\mathbf{x}}) = \sum_{r=0}^{m} \log\left(\frac{p(y_r|x_r)}{p(y_r)}\right) - k, \tag{6.59}$$

$$L_n(\tilde{\mathbf{x}}') = \sum_{r=0}^{n} \log\left(\frac{p(y_r|x_r')}{p(y_r)}\right) - k, \tag{6.60}$$

into the expression (6.58) and use $\alpha = 1/(1+\rho)$. We rewrite the exponentials in (6.58) as

$$2^{kn\rho}\overline{\int_{\mathbf{y}} p(\mathbf{y}|\mathbf{x}) \exp\left(-\alpha\rho L_m(\tilde{\mathbf{x}})\right) \left(\overline{\exp\left(\alpha L_n(\tilde{\mathbf{x}}')\right)}\right)^{\rho} d\mathbf{y}}$$

$$= \begin{cases} 2^{-(m-n)E_c(\rho)-m(E_{ce}(\rho)-k\rho)} & \text{if } m \geq n, \\ 2^{-(n-m)(E_e(\rho)-k\rho)-n(E_{ce}(\rho)-k\rho)} & \text{if } n \geq m. \end{cases} \tag{6.61}$$

The exponents used are given by

$$2^{-E_c(\rho)} = \int_y \sum_x q(x)p(y|x) \left(\frac{p(y|x)}{p(y)}2^{-k}\right)^{-\rho/(1+\rho)} dy$$

$$= 2^{k\rho/(1+\rho)} \int_y \sum_x q(x)p(y|x)^{1/(1+\rho)} p(y)^{\rho/(1+\rho)} dy$$

$$\leq 2^{k\rho/(1+\rho)} \left(\int_y \left(\sum_x q(x)p(y|x)^{1/(1+\rho)}\right)^{1+\rho} dy\right)^{1/(1+\rho)}$$

$$= 2^{k\rho/(1+\rho)-E_0(\rho,\mathbf{q})/(1+\rho)} = f_c, \tag{6.62}$$

where we have used Hölder's inequality (see Chapter 5, page 144) with $\beta_i = \sum_x q(x)p(y|x)^{1/(1+\rho)}$, $\gamma_i = p(y)^{\rho/(1+\rho)}$, and $\lambda = 1/(1+\rho)$, and "magically" there appears the error exponent from equation (5.51)! Analogously, we also find

$$2^{-(E_e(\rho)-k\rho)} = 2^{k\rho} \int_y p(y) \left(\sum_{x'} q(x') \left(\frac{p(y|x')}{p(y)}2^{-k}\right)^{1/(1+\rho)}\right)^{\rho} dy$$

$$= 2^{k\rho^2/(1+\rho)} \int_y p(y)^{1/(1+\rho)} \left(\sum_{x'} q(x')p(y|x')^{1/(1+\rho)}\right)^{\rho} dy$$

$$\leq 2^{k\rho^2/(1+\rho)} \left(\int_y \left(\sum_{x'} q(x')p(y|x')^{1/(1+\rho)}\right)^{1+\rho} dy\right)^{\rho/(1+\rho)}$$

$$= 2^{k\rho^2/(1+\rho)-E_0(\rho,\mathbf{q})/(1+\rho)} = f_e, \tag{6.63}$$

where $\lambda = \rho/(1 + \rho)$. Finally we obtain

$$
\begin{aligned}
2^{-(E_{ce}(\rho)-k\rho)} &= 2^{k\rho} \int_y \sum_x q(x)p(y|x) \left(\sum_{x'} q(x') \left(\frac{p(y|x')}{p(y|x)} \right)^{1/(1+\rho)} \right)^{\rho} dy \\
&= 2^{k\rho} \int_y \sum_x q(x)p(y|x)^{1/(1+\rho)} \left(\sum_{x'} q(x')p(y|x')^{1/(1+\rho)} \right)^{\rho} dy \\
&= 2^{k\rho - E_0(\rho,\mathbf{q})}.
\end{aligned}
\tag{6.64}
$$

Note now that, since $1 = \rho/(1 + \rho) + 1/(1 + \rho)$, we have

$$
2^{k\rho - E_0(\rho,\mathbf{q})} = 2^{k\rho^2/(1+\rho)-E_0(\rho,\mathbf{q})/(1+\rho)} 2^{k\rho/(1+\rho)-E_0(\rho,\mathbf{q})/(1+\rho)} = f_e f_c \tag{6.65}
$$

where f_e and f_c are defined in (6.62) and (6.63).

With this we can rewrite (6.58) as

$$
\text{Avg}\{\Pr(C_0 \geq N_c)\} \leq N_c^{-\rho} \sum_c p(c) \sum_{m=0}^{\infty} \sum_{n=0}^{\infty} f_e^n f_c^m. \tag{6.66}
$$

The double infinite sum in (6.66) converges if $f_e, f_c < 1$ and, hence from (6.65), if $\rho k < E_0(\rho, \mathbf{q})$, and we obtain

$$
\text{Avg}\{\Pr(C_0 \geq N_c)\} \leq N_c^{-\rho} \frac{1}{(1 - f_e)(1 - f_c)}. \tag{6.67}
$$

Similarly, it can be shown [30] that there exists a lower bound on the number of computations, given by

$$
\text{Avg}\{\Pr(C_j \geq N_c)\} \geq N_c^{-\rho}(1 - o(N_c)) \tag{6.68}
$$

Together, (6.67) and (6.68) characterize the computational behavior of sequential decoding. If $\rho \leq 1$, the expectation of (6.67) and (6.68) (i.e., the expected number of computations) becomes unbounded since

$$
\sum_{N_c=1}^{\infty} N_c^{-\rho} \tag{6.69}
$$

diverges for $\rho \leq 1$ or $k \geq R_0$. Information theory therefore tells us that we cannot beat the capacity limit by using very powerful codes and resorting to sequential decoding, since what happens is that as soon as the code rate reaches R_0 the expected number of computations per node tends to infinity. In effect our decoder fails through buffer overflow. This is why R_0 is often referred to as the *computational cutoff rate*.

Further credence to R_0 is given by the observation that rates $R = R_0$ at bit error probabilities of $P_b = 10^{-5}$–10^{-6} can be achieved with trellis codes.

This observation was made by Wang and Costello [34], who constructed random trellis codes for 8-PSK and 16-QAM constellations that achieve R_0 with constraint lengths of 15 and 16 (very realizable codes).

6.8 SOME FINAL REMARKS

As we have seen there are two broad classes of decoders, the depth-first and the breadth-first algorithms. Many attempts have been made at comparing the respective properties of these two basic approaches; for example [9] or, for convolutional codes, [30] are excellent and inexhaustible sources of information. Many of the random coding arguments in [30] for convolutional codes can be extended to trellis codes with little effort.

Where are we standing then? Sequential decoding has been popular in particular for relatively slow transmission speeds, since the buffer sizes can then be dimensioned such that buffer overflow is controllable. Sequential decoding, however, suffers from two major drawbacks. First, it is a "sequential" algorithm; that is, modern pipelining and parallelizing is very difficult, if not impossible, to accomplish. Second, the metric used in sequential decoding contains the "bias" term accounting for the different path lengths. This makes sequential decoding highly channel dependent. Furthermore, this bias term may be prohibitively complex to calculate for other than straight channel coding applications (see, e.g., [35]).

Breadth-first search algorithms, in particular the optimal Viterbi algorithm and the popular M-algorithm, do not suffer from the metric "bias" term. These structures can also be parallelized much more readily, which makes them good candidates for VLSI implementations. They are therefore very popular for high-speed transmission systems. The Viterbi algorithm can be implemented with a separate metric calculator for each state. More on the implementation aspects of parallel Viterbi decoder structures can be found in [31–37]. The Viterbi decoder has proven so successful in applications that it is the algorithm of choice for most applications of code decoding at present.

The M-algorithm can also be implemented by exploiting inherent parallelism of the algorithm, and [38] discusses an interesting implementation that avoids the sorting of the paths associated with the basic algorithm. The M-algorithm has also been successfully applied to multiuser detection, a problem that can also be stated as a trellis (tree) search [39], and to the decoding of block codes. This latter subject will be discussed further in Section 7.9.

APPENDIX

In this appendix we calculate the vector Euclidean distance for a specific set C_p of retained paths. Noting that δ_n from (6.20) is a vector of Gaussian random variables (see also Chapter 2), we can easily write its probability density function [26], viz.,

$$p(\delta_n) = \frac{1}{(2\pi)^{M/2}|\mathbf{R}|^{1/2}} \exp\left(-\tfrac{1}{2}(\mu - \delta_n)^T \mathbf{R}^{-1}(\mu - \delta_n)\right), \qquad (6.70)$$

where μ is the vector of mean values given by $\mu_i = d_i^2$, and $\mathbf{R}$ is the covariance matrix of the Gaussian random variables δ_n whose entries $r_{ij} = \mathrm{E}\left[(\delta_n^{(i,c)} - \mu_i)(\delta_n^{(j,c)} - \mu_j)\right]$ can be evaluated as

$$r_{ij} = \begin{cases} 2N_0(d_i^2 + d_j^2 - d_{ij}^2) & \text{if } i \neq j, \\ 4N_0 d_i^2 & \text{if } i = j, \end{cases} \qquad (6.71)$$

and where

$$d_{ij}^2 = \left|\tilde{\mathbf{x}}^{(p_i)} - \tilde{\mathbf{x}}^{(p_j)}\right|^2 \qquad \text{and} \qquad d_i^2 = \left|\tilde{\mathbf{x}}^{(p_i)} - \tilde{\mathbf{x}}^{(c)}\right|. \qquad (6.72)$$

The vector μ of mean values is given by $\mu_i = d_i^2$.

Now the probability of losing the correct path at time n can be calculated by

$$\Pr(\mathrm{CPL}|C_p) = \int_{\delta_n \leq 0} p(\delta_n)\, d\delta_n. \qquad (6.73)$$

Equation (6.73) is difficult to evaluate due to the correlation of the entries in δ_n, but one thing we know is that the area $\delta_n \leq \mathbf{0}$ of integration is convex. This allows us to place a hyperplane through the point closest to the center of the probability density function, μ and to bound (6.73) above by the probability that the noise carries the point μ across this hyperplane. This results in a simple one-dimensional integral, whose value is given by (compare also (2.14))

$$\Pr(\mathrm{CPL}|C_p) \leq Q\left(\sqrt{\frac{d_l^2}{2N_0}}\right), \qquad (6.21)$$

where d_l^2, the vector Euclidean distance, is

$$d_l^2 = 2N_0 \min_{\mathbf{y} \leq 0}(\mu - \mathbf{y})^T \mathbf{R}^{-1}(\mu - \mathbf{y}), \qquad (6.74)$$

and $\mathbf{y}$ is here simply a dummy variable of minimization.

The problem of calculating (6.21) has now been transformed into the geometric problem of finding the point on the surface of the convex polytope $\mathbf{y} \leq \mathbf{0}$ that is closest to μ using the distance measure of (6.74). This situation is illustrated in Figure 6.14 for a two-dimensional scenario. The minimization in (6.74) is a constrained minimization of a quadratic form. Obviously, some of the constraints $\mathbf{y} \leq \mathbf{0}$ will be met with equality. These constraints are called the *active constraints*; that is, if $\mathbf{y} = (\mathbf{y}^{(a)}, \mathbf{y}^{(p)})^T$ is the partitioning of $\mathbf{y}$ into active and passive components, $\mathbf{y}^{(a)} = \mathbf{0}$. This minimum is the point $\mathbf{y}_0$ in Figure 6.14. The right side of Figure 6.14 also shows the geometric configuration when the decorrelating linear transformation $\delta'_n = \sqrt{\mathbf{R}^{(-1)}} \delta_n$ is applied. The vector Euclidean distance (6.74) is invariant to such a transformation, but $E\left[(\delta_n'^{(i,c)} - \mu'_i)(\delta_n'^{(j,c)} - \mu'_j) \right] = \delta_{ij}$; i.e., the decorrelated metric differences are independent with unit variance each. Naturally we may work in either space. Since the random variables δ'_n are independent, equal-variance Gaussian, we know from basic communication theory [26, Chapter 2], that the probability that μ' is carried into the shaded region of integration can be bounded above by integrating over the half-plane not containing μ', as illustrated in the figure. This leads to (6.21).

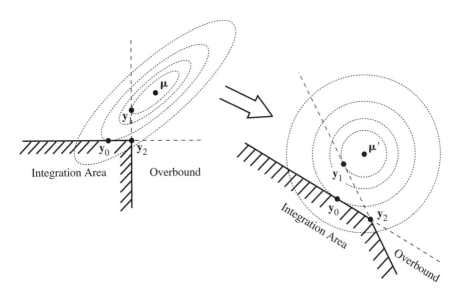

Figure 6.14 Illustration of the concept of the vector Euclidean distance with $M = 2$. The distance between $\mathbf{y}_0$ and μ is d_l^2. The right side shows the space after decorrelation, and d_l^2 equals the standard Euclidean distance between $\mathbf{y}'_0$ and μ'.

We now have to minimize

$$d_l^2 = 2N_0 \min_{\mathbf{y}^{(p)} \leq 0} \left(\begin{pmatrix} \boldsymbol{\mu}^{(p)} \\ \boldsymbol{\mu}^{(a)} \end{pmatrix} - \begin{pmatrix} \mathbf{y}^{(p)} \\ \mathbf{0} \end{pmatrix} \right)^T \begin{pmatrix} \mathbf{R}^{(pp)} & \mathbf{R}^{(pa)} \\ \mathbf{R}^{(ap)} & \mathbf{R}^{(aa)} \end{pmatrix}^{-1} \left(\begin{pmatrix} \boldsymbol{\mu}^{(p)} \\ \boldsymbol{\mu}^{(a)} \end{pmatrix} - \begin{pmatrix} \mathbf{y}^{(p)} \\ \mathbf{0} \end{pmatrix} \right),$$
(6.75)

where we have partitioned $\boldsymbol{\mu}$ and $\mathbf{R}$ analogously to $\mathbf{y}$. After some elementary operations we obtain

$$\mathbf{y}^{(p)} = \boldsymbol{\mu}^{(p)} + \left(\mathbf{X}^{(pp)} \right)^{-1} \mathbf{X}^{(pa)} \boldsymbol{\mu}^{(a)} \leq \mathbf{0},$$
(6.76)

and

$$d_l^2 = 2N_0 \boldsymbol{\mu}^{(a)T} \left(\mathbf{X}^{(aa)} \right)^{-1} \boldsymbol{\mu}^{(a)},$$
(6.77)

where[6]

$$\mathbf{X}^{(pp)} = \left[\mathbf{R}^{(pp)} - \mathbf{R}^{(pa)} \left(\mathbf{R}^{(aa)} \right)^{-1} \mathbf{R}^{(ap)} \right]^{-1},$$

$$\mathbf{X}^{(aa)} = \left(\mathbf{R}^{(aa)} \right)^{-1} \left[\mathbf{I} + \mathbf{R}^{(ap)} \mathbf{X}^{(pp)} \mathbf{R}^{(pa)} \right],$$

$$\mathbf{X}^{(pa)} = -\mathbf{X}^{(pp)} \mathbf{R}^{(pa)} \left(\mathbf{R}^{(aa)} \right)^{-1}.$$
(6.78)

We are now presented with the problem of finding the active components in order to evaluate (6.77). This is a combinatorial problem; in other words, we must test all $2^M - 1 = \sum_{i=1}^{M} \binom{M}{i}$ possible combinations of active components from the M entries in $\mathbf{y}$ for compatibility with (6.76). This gives us the following procedure:

Step 1.   Select all $2^M - 1$ combinations of active components and set $\mathbf{y}^{(a)} = \mathbf{0}$ for each.

Step 2.   For each combination for which $\mathbf{y}^{(p)} \leq \mathbf{0}$, store the resulting d_l^2 from (6.77) in a list.

Step 3.   Select the smallest entry from the list in Step 2 as d_l^2.

As an example, consider again Figure 6.14. The $2^2 - 1 = 3$ combinations correspond to the points $\mathbf{y}_0$, $\mathbf{y}_1$ and $\mathbf{y}_2$. The point $\mathbf{y}_1$ does not qualify,

---

[6] These equations can readily be derived from the partitioned matrix inversion lemma:

$$\begin{pmatrix} A & B \\ C & D \end{pmatrix}^{-1} = \begin{pmatrix} E & F \\ G & H \end{pmatrix},$$

where

$$E = \left(A - BD^{-1}C \right)^{-1}, \quad F = -EBD^{-1}, \quad G = -D^{-1}CE,$$

and

$$H = D^{-1} + D^{-1}CEBD^{-1}.$$

because it violates (6.76). The minimum is chosen between $\mathbf{y}_0$ and $\mathbf{y}_2$. This process might be easier to visualize in the decorrelated space $\mathbf{y}'$, where all the distances are ordinary Euclidean distances and the minimization becomes obvious.

One additional complication needs to be addressed. The correlation matrix $\mathbf{R}$ may be singular. This happens when one or more entries in $\mathbf{y}$ are linearly dependent on the other entries. In the context of the restriction $\mathbf{y} \leq \mathbf{0}$, we have redundant conditions. The problem, again, is that of finding the redundant entries that can be dropped from consideration. Fortunately, our combinatorial search helps us here. Since we are examining all combinations of possible active components, we may simply drop any dependent combinations that produce a singular $\mathbf{R}^{(aa)}$ from further consideration without affecting d_l^2.

REFERENCES

[1] R. M. Fano, "A heuristic discussion of probabilistic decoding," *IEEE Trans. Inform. Theory*, Vol. IT-9, pp. 64–74, 1963.

[2] J. L. Massey, "Variable-length codes and the Fano metric," *IEEE Trans. Inform. Theory*, Vol. IT-18, pp. 196–198, 1972.

[3] J. M. Wozencraft and B. Reiffen, *Sequential Decoding*, MIT Press, Cambridge, MA, 1961.

[4] K. Sh. Zigangirov, "Some sequential decoding procedures," *Prob. Pederachi Inform.*, Vol. 2, pp. 13–25, 1966.

[5] F. Jelinek, "A fast sequential decoding algorithm using a stack," *IBM J. Res. Dev.*, Vol. 13, pp. 675–685, 1969.

[6] S. Lin and D. J. Costello, Jr., *Error Control Coding*, Prentice-Hall, Englewood Cliffs, NJ, 1983.

[7] J. B. Anderson and S. Mohan, "Sequential coding algorithms: A survey and cost analysis," *IEEE Trans. Commun.*, Vol. COM-32, No. 2, pp. 169–176, 1984.

[8] J. B. Anderson and S. Mohan, *Source and Channel Coding: An Algorithmic Approach*, Kluwer, Boston, 1991.

[9] G. J. Pottie and D. P. Taylor, "A comparison of reduced complexity decoding algorithms for trellis codes," *IEEE J. Select. Areas Commun.*, Vol. SAC-7, No. 9, pp. 1369–1380, 1989.

[10] D. Haccoun and M. J. Ferguson, "Generalized stack algorithms for decoding convolutional codes," *IEEE Trans. Inform. Theory*, Vol. IT-21, pp. 638–651, 1975.

[11] P. R. Chevillat and D. J. Costello, Jr., "A multiple stack algorithm for erasure-free decoding of convolutional codes," *IEEE Trans. Commun.*, Vol. COM-25, pp. 1460–1470, 1977.

[12] G. J. Foschini, "A reduced state variant of maximum likelihood sequence detection attaining optimum performance for high signal-to-noise ratios," *IEEE Trans. Inform. Theory*, Vol. IT-23, pp. 605–609, 1977.

[13] J. M. Geist, "An empirical comparison of two sequential decoding algorithms," *IEEE Trans. Commun.*, Vol. COM-19, pp. 415–419, 1971.

[14] J. M. Geist, "Some properties of sequential decoding algorithms," *IEEE Trans. Inform. Theory*, Vol. IT-19, pp. 519–526, 1973.

[15] T. Aulin, "Breadth first maximum likelihood sequence estimation," unpublished manuscript.

[16] T. Aulin, "Recovery Properties of the SA(B) Algorithm," Technical Report No. 105, Chalmers University of Technology, Sweden, February 1991.

[17] L. Ma, "Suboptimal decoding strategies," MSEE thesis, University of Texas at San Antonio, May 1996.

[18] H. Osthoff, J. B. Anderson, R. Johannesson, and C. Lin, "Systematic feed-forward convolutional encoders are better than other encoders with an M-algorithm decoder," unpublished manuscript, 1995.

[19] K. S. Zigangirov and V. D. Kolesnik, "List decoding of trellis codes," *Prob. Control Inform. Theory*, No. 6, 1980.

[20] S. J. Simmons and P. Wittke, "Low complexity decoders for constant envelope digital modulation," *Conf. Rec.*, GlobeCom, Miami, FL, pp. E7.7.1–E7.7.5, November 1982.

[21] T. Aulin, "Study of a new trellis decoding algorithm and its applications," Final Report, ESTEC Contract 6039/84/NL/DG, European Space Agency, Noordwijk, The Netherlands, December 1985.

[22] K. Balachandran, "Design and performance of constant envelope and non-constant envelope digital phase modulation schemes," Ph.D. thesis, ECSE Dept. Renselear Polytechnic Institute, Troy, NY, February 1992.

[23] A. J. Viterbi, "Error bounds for convolutional codes and an asymptotically optimum decoding algorithm," *IEEE Trans. Inform. Theory*, Vol. IT-13, pp. 260–269, 1969.

[24] J. K. Omura, "On the Viterbi decoding algorithm," *IEEE Trans. Inform. Theory*, Vol. IT-15, pp. 177–179, 1969.

[25] G. D. Forney, Jr., "The Viterbi algorithm," *Proc. IEEE*, Vol. 61, pp. 268–278, 1973.

[26] J. G. Proakis, *Digital Communications,* McGraw-Hill, New York, 1989.

[27] G. D. Forney, "Maximum-likelihood sequence estimation of digital sequences in the presence of intersymbol interference," *IEEE Trans. Inform. Theory*, Vol. IT-18, pp. 363–378, 1972.

[28] S. Verdú, "Minimum probability of error for asynchronous Gaussian multiple-access channels," *IEEE Trans. Inform. Theory*, Vol. IT-32, pp. 85–96, 1986.

[29] R. E. Blahut, *Principles and Practice of Information Theory*, Addison-Wesley, Reading, MA, 1987.

[30] A. J. Viterbi and J. K. Omura, *Principles of Digital Communication and Coding*, McGraw-Hill, New York, 1979.

[31] G. C. Clark and J. B. Cain, *Error-Correction Coding for Digital Communications*, Plenum Press, New York, 1983.

[32] R. J. McEliece, "On the BCJR trellis for linear block codes," *IEEE Trans. Inform. Theory*, Vol. IT-42, No. 4, pp. 1072–1092, 1996.

[33] L. R. Bahl, J. Cocke, F. Jelinek, and J. Raviv, "Optimal decoding of linear codes for minimizing symbol error rate," *IEEE Trans. Inform. Theory*, Vol. IT-20, pp. 284–287, 1974.

[34] F.-Q. Wang and D. J. Costello, Jr., "Probabilistic construction of large constraint length trellis codes for sequential decoding," *IEEE Trans. Commun.*, Vol. COM-43, No. 9, pp. 2439–2448, 1995.

[35] L. Wei, L. K. Rasmussen, and R. Wyrwas, "Near optimum tree-search detection schemes for bit-synchronous CDMA systems over Gaussian and two-path rayleigh fading channels," unpublished manuscript.

[36] G. Feygin and P. G. Gulak, "Architectural tradeoffs for survivor sequence memory management in Viterbi decoders," *IEEE Trans. Commun.*, Vol. COM-41, pp. 425–429, 1993.

[37] O. M. Collins, "The subtleties and intricacies of building a constraint length 15 convolutional decoder," *IEEE Trans. Commun.*, Vol. COM-40, pp. 1810–1819, 1992.

[38] S. J. Simmons, "A nonsorting VLSI structure for implementing the (M,L) algorithm," *IEEE J. Select. Areas Commun.*, Vol. SAC-6, pp. 538–546, 1988.

[39] L. Wei and C. Schlegel, "Synchronous DS-SSMA with improved decorrelating decision-feedback multiuser detection," *IEEE Trans. Veh. Technol.*, Vol. VT-43, No. 3, pp. 767–772, 1994.

CHAPTER 7

LINK TO BLOCK CODES

7.1 PRELIMINARIES

This book is mostly about trellis codes. Although trellis codes have almost universally been used in error control systems for memoryless channels, we must not brush aside block codes. Indeed, most theoretical work in coding theory, starting with Shannon, has been done with block codes. As a consequence, the majority of theoretical results known in coding theory are for block codes. Trellis codes have mostly been used for channel coding, whereas block codes have been used extensively in important applications, most notably in compact disk (CD) storage, digital audio taping (DAT), and high-definition television (HDTV) [1]. Arguably, the single most important result that accounts for the use and popularity of block codes in applications is the algebraic decoding methods for the general class of Goppa codes [2], in particular the Berlekamp-Massey algorithm [4, 5]. Reed-Solomon codes, and to a lesser extent Bose-Chaudhuri-Hocquenghem (BCH) codes, are the most popular subclasses of Goppa codes. These codes can be decoded very fast with a complexity on the order of $O(d^2)$, where d is the minimum distance of the code, using finite-field arithmetic that can be implemented efficiently in very large scale integrated (VLSI) circuits. However, these efficient algebraic decoding algorithms only perform error correction, and soft-channel information cannot be used. This may explain why the majority of block code applications have been for systems requiring error correction. Soft-decision decoding of block codes can be done, however, and several algorithms have been proposed over the last three decades [6–9], but have never found much attention by practitioners of coding theory. More recently, soft-decision decoding of block codes has seen a revival of interest [10–18], in particular in connection with the trellis complexity problem of block codes addressed in this chapter.

In Chapter 6 we discussed a number of powerful decoding techniques that use the inherent trellis structure of the codes. This made it logical to use this trellis in the decoder. Block codes, on the other hand, have traditionally been constructed and decoded in algebraic ways. Algebraic decoding does not lend itself easily to include soft decisions, as we have done in decoding trellis codes. We will see in this chapter that block codes can also be described by a code trellis; therefore, all the decoding algorithms from the previous chapter find application. We will further see that block codes can also be constructed from the trellis approach and that families of long known block codes reappear in a different light. The results discussed in this chapter, tree- and trellis-based soft-decision decoding of block codes, have spurred renewed interest in block codes and bring block and trellis codes much closer together.

Block codes are extensively discussed in a number of excellent text-books, among them [3, 2, 19–21], where the book by MacWilliams and Sloane [3] is arguably the most comprehensive and thorough treatment of block codes to date. It is not the intention of this chapter to discuss block codes at any length but to establish the connection between trellis and block codes. To this end we need some basic block coding concepts.

7.2 BLOCK CODE PRIMER

One of the easiest ways to define a linear[1] block code of rate $R = k/n$ is via its *parity-check matrix* $\mathbf{H} = [\mathbf{h}_1, \mathbf{h}_2, \ldots, \mathbf{h}_n]$, which is an $n - k \times n$ matrix where $\mathbf{h}_j$ is the jth column and the entries $h_{ij} \in \mathrm{GF}(p)$, where $\mathrm{GF}(p)$ is the finite field of the integers modulo p and p is a prime. In the simplest case the code extends over $\mathrm{GF}(2)$, i.e., over $\{0, 1\}$, and all operations are modulo 2, or XOR and AND operations.

A linear block code may now be characterized by the following

DEFINITION 7.1

A linear block code $\mathcal{C}$ with parity check matrix $\mathbf{H}$ is the set of all n-tuples (vectors), or code words $\mathbf{x}$, such that

$$\mathbf{Hx} = 0. \tag{7.1}$$

For example, the family of single-error-correcting binary Hamming codes has the generator matrices $\mathbf{H}$ whose columns are all the $2^m - 1$

---

[1] We restrict attention to linear block codes since they are the largest and most important class of block codes.

nonzero binary vectors of length $m = n - k$. These codes have a rate $R = k/n = (2^m - m - 1)/(2^m - 1)$ and a minimum distance $d_{\min} = 3$ [3].

The first such code found by Richard Hamming in 1948 has the parity-check matrix

$$\mathbf{H}_{[7,4]} = \begin{bmatrix} 1 & 1 & 1 & 0 & 1 & 0 & 0 \\ 1 & 1 & 0 & 1 & 0 & 1 & 0 \\ 1 & 0 & 1 & 1 & 0 & 0 & 1 \end{bmatrix} = [\mathbf{h}_1\ \mathbf{h}_2\ \mathbf{h}_3\ \mathbf{h}_4\ \mathbf{h}_5\ \mathbf{h}_6\ \mathbf{h}_7]. \qquad (7.2)$$

Note that each row of $\mathbf{H}$ corresponds to a parity-check equation. The parity-check matrix can always be arranged such that $\mathbf{H} = [\mathbf{A}|\mathbf{I}_{n-k}]$ through column and row permutations and linear combinations, where $\mathbf{I}_{n-k}$ is the $n - k$ identity matrix and $\mathbf{A}$ has dimensions $n - k \times k$. This has already been done in (7.2). From this form we can obtain the systematic $k \times n$ *code generator matrix*

$$\mathbf{G} = \left[\mathbf{I}_k \middle| \left(-\mathbf{A}^T\right)\right] \qquad (7.3)$$

through simple matrix manipulations. The code generator matrix has dimensions $k \times n$ and is used to generate directly the code words $\mathbf{x}$ via $\mathbf{x}^T = \mathbf{u}^T\mathbf{G}$, where $\mathbf{u}$ is the information k-tuple.

7.3   TRELLIS DESCRIPTION OF BLOCK CODES

In this section we use a trellis as a visual method of keeping track of the p^k codewords of a block code $\mathcal{C}$, and each distinct codeword corresponds to a distinct path through the trellis. This idea was first explored by Bahl et al. [22] in 1974, and later also by Wolf [23] and Massey [24] in 1978. Following the approach in [22, 23] we define the states of the trellis as follows: Let s_r be the label of the state at time r. Then

$$s_{r+1} = s_r + x_r\mathbf{h}_r = \sum_{l=1}^{r} x_l\mathbf{h}_l, \qquad (7.4)$$

where x_r runs through all permissible code symbols at time r. The state at the end of time interval r, s_{r+1}, is calculated from the preceding state s_r according to (7.4). If the states s_r and s_{r+1} are connected, they are joined in the trellis diagram by a branch labeled with the output symbol x_r that caused the connection. We see that (7.4) simply implements the parity-check equation (7.1) in a recursive fashion, since the final zero state $s_{n+1} = \sum_{l=1}^{n} x_l\mathbf{h}_l = \mathbf{Hx}$ is the complete parity-check equation! Since each intermediate state $s_r = \sum_{l=1}^{r-1} x_l\mathbf{h}_l$ is a vector of length $n - k$ with elements in GF(p), also called the *partial syndrome* at time r, there can be at most

p^{n-k} distinct states at time r, since that is the maximum number of distinct p-ary vectors of length $n - k$. We will see that often the number of states is significantly less than that.

At this point an example seems appropriate. Let us construct the trellis of the [7, 4] Hamming code, whose parity-check matrix is given by (7.2). The corresponding trellis is shown in Figure 7.1. The states s_r are labeled as ternary vectors $(\sigma_1, \sigma_2, \sigma_3)^T$ in accordance with (7.4).

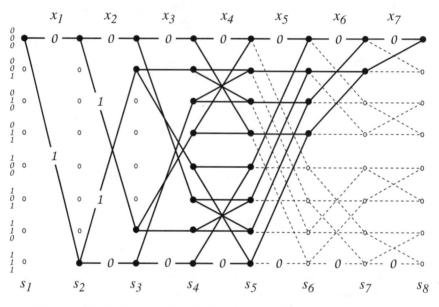

Figure 7.1 Trellis diagram of the [7, 4] Hamming code discussed in the text. The labeling of the trellis is such that a 0 causes a horizontal transition and a 1 causes a sloped transition. This follows from (7.4) since $x_r = 0$ causes $s_{r+1} = s_r$. Hence, only a few transitions are labeled in the figure.

Note that all the path extensions that lead to states $s_{n+1} \neq (000)^T$ are dashed, since $s_n^{(i)}$ is the final syndrome and must equal $(000)^T$ for $\mathbf{x}$ to be a codeword. We therefore force the trellis to terminate in the zero state at time n.

Some further remarks are in order. Contrary to the trellis codes discussed earlier, the trellis of block codes is time varying; i.e., the connections are different for each section of the trellis. The number of states is also varying, with a maximum number of states smaller than or equal to p^{n-k}. The codes are regular in the sense of Section 3.3; that is, the error probability is independent of the chosen correct path. This follows directly from the linearity of the block codes considered.

It becomes evident that using a trellis decoding algorithm is quite a different game from algebraic decoding, which operates on the syndrome of the entire codeword. There the syndrome for some received symbol sequence $\mathbf{y}$, $\mathbf{s} = \mathbf{Hy}$, is calculated, from which the error pattern $\mathbf{e}$ is estimated [3]. Then the hypothesized $\hat{\mathbf{x}} = \mathbf{y} - \mathbf{e}$ is generated from y and, in a final step, $\hat{\mathbf{u}}$ is decoded from $\hat{\mathbf{x}}$. Trellis coding operates quite differently, and we see that, starting at step $r = 4$ in Figure 7.1, hypotheses are being discarded through merger elimination, even before the entire vector y is received. This entails no loss since we know from Chapter 6 that this procedure is optimal.

7.4   MINIMAL TRELLISES

In this section we explore the question of optimality of the trellis representation of block codes introduced in the last section. Following the results of McEliece [25], we will show that the parity-check (PC) trellis representation from Section 7.3 is the best possible in the sense that it minimizes the number of states as well as the number of edges in the trellis for each time instant r. Since the number of edges is the relevant measure of complexity for the Viterbi algorithm (Theorem 6.2), the PC trellis also minimizes the decoding complexity if Viterbi decoding is used.

From (7.4) we know that the set of states at time r, denoted by $\mathcal{S}_r$, is given by the mapping $\mathcal{C} \mapsto \mathcal{S}_r$:

$$\{s_r(x) : s_r(x) = x_1\mathbf{h}_1 + \cdots + x_{r-1}\mathbf{h}_{r-1}\}. \tag{7.5}$$

This mapping is linear due to the linearity of the code and may be viewed as a vector space. The maximum dimension of this vector space is $n - k$, since this is the length of the vectors $s_r(x)$, but is, in general, much less (see, e.g., Figure 7.1). We denote its dimension by σ_r ($\sigma_r = \dim \mathcal{S}_r$). Note that for binary codes the dimension σ_r of the state-space implies that the number of states $|\mathcal{S}_r|$ in the trellis at time r is 2^{σ_r}.

Likewise, an edge in the trellis is specified by the state from where it originates, the state into which it leads, and the symbol with which it is labeled. Formally we may describe the edge space $\mathcal{B}_{r,r+1}$ at time r by

$$\{b_{r,r+1}(x) : b_{r,r+1}(x) = (s_r(x), s_{r+1}(x), x_r)\}. \tag{7.6}$$

Clearly, the edge mapping is also linear, and we denote its dimension by $\beta_{r,r+1}$.

Now, since codewords are formed via $\mathbf{x} = \mathbf{G}^T \mathbf{u}$,

$$s_r(x) = \mathbf{H}_r(x_1, \ldots, x_{r-1}) = \mathbf{H}_r \mathbf{G}_r^T \mathbf{u}, \tag{7.7}$$

where $\mathbf{H}_r$ and $\mathbf{G}_r$ are the matrices made up of the first $r - 1$ columns of $\mathbf{H}$ and $\mathbf{G}$, respectively. So, $\mathbf{H}_r \mathbf{G}_r^T$ is a $(r - 1) \times k$ matrix, and we have

LEMMA 7.1

The state-space $\mathcal{S}_r$ at the completion of time r is the column-space of $\mathbf{H}_r \mathbf{G}_r^T$, and, consequently, its dimension σ_r is the rank of the matrix $\mathbf{H}_r \mathbf{G}_r^T$.

Note that we can, of course, span the trellis up backwards from the end and then

$$s_r(x) = \overline{\mathbf{H}}_r(x_r, \ldots, x_n)^T = \overline{\mathbf{H}}_r \overline{\mathbf{G}}_r^T u, \tag{7.8}$$

where $\overline{\mathbf{H}}_r^T$ and $\overline{\mathbf{G}}_r$ are matrices made up of the last $n - (r - 1)$ columns of $\mathbf{H}$ and $\mathbf{G}$, respectively. $\overline{\mathbf{H}}_r \overline{\mathbf{G}}_r^T$ is a $n - (r - 1) \times k$ matrix, and, using the fact that the rank of a matrix is smaller or equal to its smallest dimension, we conclude from (7.7) and (7.8) and the last section that, according to

LEMMA 7.2

The dimension σ_r of the state-space $\mathcal{S}_r$ at the completion of time $(r - 1)$ is bounded by

$$\sigma_r \leq \min((r - 1), n - (r - 1), k, n - k). \tag{7.9}$$

Lemma 7.9 can be seen nicely in Figure 7.1.

Now, consider all the codewords or code sequences for which $x_j = 0, j > r$. These sequences are called the past subcode, denoted by P_r, and are illustrated in Figure 7.2 for $r = 5$ for the PC trellis for the [7, 4] Hamming code. We denote the dimension of P_r by p_r. Likewise, we define the future subcode F_r as the set of code sequences for which $x_j = 0, j \leq r$, and denote its dimension by f_r. The future subcode F_2 for the [7, 4] Hamming code is also shown in Figure 7.2. Clearly, both of these subcodes are linear subcodes of the original code; that is, they are subgroups of the original group (code) $\mathcal{C}$.

Now, mathematically, $s_r(x)$ is a map from the code space $\mathcal{C}$, which is k-dimensional, into the state-space $\mathcal{S}_r$ at time r, which is σ_r-dimensional. Such a linear map has a kernel, which is the set of codewords mapped into the state $s_r(x) = 0$ (i.e., all the codewords whose trellis paths pass through the zero state at time r). From Figure 7.2 we now guess the following

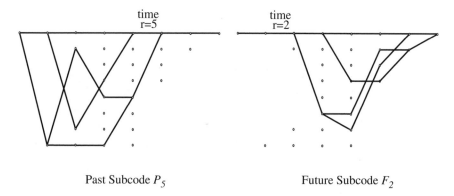

Past Subcode P_5 Future Subcode F_2

Figure 7.2 Trellis diagram of the past subcode P_5 and the future subcode F_2 of the [7, 4] Hamming code whose full trellis is shown in Figure 7.1.

LEMMA 7.3

The kernel of the map $s_r(x)$ is the sum of the past and future subcodes P_r and F_r:

$$\text{Ker}(s_r) = P_{r-1} \oplus F_{r-1}. \tag{7.10}$$

The kernel $\text{Ker}(s_r)$ is illustrated in Figure 7.3.

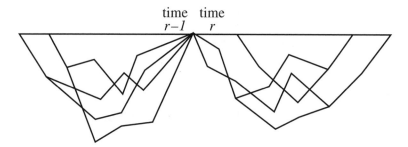

Figure 7.3 Illustration of the kernel of the map $\mathcal{C} \mapsto \mathcal{S}_r$ as the set of codewords that pass through $s_r(x) = 0$.

Proof

If $x \in P_{r-1} \oplus F_{r-1}$, it can be expressed as $\mathbf{x} = \mathbf{x}^{(1)} + \mathbf{x}^{(2)}$, where $\mathbf{x}^{(1)} \in P_{r-1}$ and $\mathbf{x}^{(2)} \in F_{r-1}$. But since $\mathbf{x}^{(1)} \in C$, $\mathbf{Hx}^{(1)} = 0$, and hence $s_r(\mathbf{x}^{(1)}) = 0$. Obviously $s_r(\mathbf{x}^{(2)}) = 0$, and since the mapping s_r is linear we conclude that $s_r(x) = 0$ and $P_{r-1} \oplus F_{r-1} \subseteq \text{Ker}(s_r)$.

Conversely, assume $s_r(\mathbf{x}) = 0$ and let $\mathbf{x} = \mathbf{x}^{(1)} + \mathbf{x}^{(2)}$ again, where we choose $\mathbf{x}^{(1)} = (x_1, \ldots, x_{r-1}, 0, \ldots, 0)$ and $\mathbf{x}^{(2)} = (0, \ldots, 0, x_r, \ldots, x_n)$. But due to $s_r(\mathbf{x}) = 0$ and (7.7), we conclude that $\mathbf{Hx}^{(1)} = 0$ and therefore $\mathbf{x}^{(1)} \in C$ and $\mathbf{x}^{(1)} \in P_{r-1}$. Likewise, applying (7.8) gives $\mathbf{Hx}^{(2)} = 0$ and $\mathbf{x}^{(2)} \in C$ and $\mathbf{x}^{(2)} \in F_{r-1}$. Hence, $\text{Ker}(s_r) \subseteq P_{r-1} \oplus F_{r-1}$ also.

As is the case with $s_r(x)$, $b_{r,r+1}(x)$ is a map from the code space C into the $\beta_{r,r+1}$-dimensional edge space. Its kernel is given by the following

LEMMA 7.4

The kernel of the map $b_{r,r+1}(x)$ is the sum of the past and future sub-codes P_{r-1} and F_r:

$$\text{Ker}(b_{r,r+1}) = P_{r-1} \oplus F_r. \tag{7.11}$$

PROOF

From the definition of the edge mapping (7.6) we see that $\text{Ker}(b_{r,r+1})$ must be contained in $\text{Ker}(s_r)$ and $\text{Ker}(s_{r+1})$, as well as obey the condition $x_r = 0$. From this the lemma follows.

The following theorem then determines the number of states and edges in the PC trellis.

THEOREM 7.5

The number of states at depth r in the PC trellis for a binary code is

$$|S_r| = 2^{k-p_{r-1}-f_{r-1}}, \tag{7.12}$$

and the number of edges at depth r is

$$|B_{r,r+1}| = 2^{k-p_{r-1}-f_r}. \tag{7.13}$$

PROOF

Since s_r is a linear map from $C \to S_r$, we may apply the rank theorem of linear algebra [26]—i.e.,

$$\dim \text{Ker}(s_r) + \dim |S_r| = \dim C \tag{7.14}$$

—which leads to $p_{r-1} + f_{r-1} + \sigma_r = k$ and hence to (7.12). The proof of (7.13) is totally analogous.

We now come to the heart of this section. The optimality of the PC trellis is asserted by

THEOREM 7.6

The number of states $|S_r|$ and the number of edges $|B_{r,r+1}|$ at depth r in any trellis that represents the linear block code C are bounded by

$$|\mathcal{S}_r| \geq 2^{k-p_{r-1}-f_{r-1}}, \tag{7.15}$$

$$|\mathcal{B}_{r,r+1}| \geq 2^{k-p_{r-1}-f_r}. \tag{7.16}$$

The PC trellis therefore simultaneously minimizes both the state and the edge count at every depth r.

Proof

The proof relies on the linearity of the code. Assume that T is a trellis that represents $\mathcal{C}$ and that $s_r(x)$ is a state at depth r in this trellis. Now let $\mathcal{C}_{s_r}$ be the subset of codewords that pass through s_r. Since every codeword must pass through at least one state at time r,

$$\mathcal{C} = \bigcup_{s_r} \mathcal{C}_{s_r}. \tag{7.17}$$

Now consider the codewords $\mathbf{x} = \begin{bmatrix} \mathbf{x}_p \ \mathbf{x}_f \end{bmatrix}^T$ in $\mathcal{C}_{s_r}$, where $\mathbf{x}_p = (x_1, \ldots, x_{r-1})^T$ is the past portion of x and $x_f = (x_r, \ldots, x_n)^T$ is the future portion. Then let $\mathbf{x}^{(1)} = \begin{bmatrix} \mathbf{x}_p^{(1)} \ \mathbf{x}_f^{(1)} \end{bmatrix}^T$ be a particular codeword in $\mathcal{C}_{s_r}$, say the one with minimum Hamming weight.

Now, since $\mathbf{x}$ and $\mathbf{x}^{(1)}$ pass through s_r, $\mathbf{x}' = \begin{bmatrix} \mathbf{x}_p \ \mathbf{x}_f^{(1)} \end{bmatrix}^T$ and $\mathbf{x}'' = \begin{bmatrix} \mathbf{x}_p^{(1)} \ \mathbf{x}_f \end{bmatrix}^T$ are two codewords that also pass through s_r. Hence,

$$\mathbf{x}' - \mathbf{x} = \begin{bmatrix} 0 \\ \mathbf{x}_f^{(1)} - \mathbf{x}_f \end{bmatrix} \in P_{r-1} \tag{7.18}$$

and

$$\mathbf{x}'' - \mathbf{x} = \begin{bmatrix} \mathbf{x}_p^{(1)} - \mathbf{x}_p \\ 0 \end{bmatrix} \in F_{r-1} \tag{7.19}$$

We conclude that

$$\mathbf{x}^{(1)} - \mathbf{x} = \begin{bmatrix} \mathbf{x}_p^{(1)} - \mathbf{x}_p \\ \mathbf{x}_f^{(1)} - \mathbf{x}_f \end{bmatrix} = \mathbf{x}' + \mathbf{x}'' - 2\mathbf{x} \Rightarrow \mathbf{x} = \mathbf{x}' + \mathbf{x}'' - \mathbf{x}^{(1)}, \tag{7.20}$$

and, hence, $\mathbf{x}$ is in the coset of $P_{r-1} \oplus F_{r-1}$ with coset leader $-\mathbf{x}^{(1)}$. Therefore $|\mathcal{C}_{s_r}| \leq 2^{p_{r-1}+f_{r-1}}$, which, together with $|\mathcal{C}| = 2^k$, immediately implies (7.15). The proof of the second part of the theorem is analogous.

The dimensions p_r and f_r can be found by inspecting the trellis, which, however, is a rather cumbersome undertaking. In addition, once the trellis is constructed the number of states and branches at each time is known, and there is no need for the values p_r and f_r anymore. There is, however, a simpler way to obtain the dimensions p_r and f_r [25]. Let us consider the example of the [7, 4] Hamming code from the last section again, whose generator matrix is

$$
\mathbf{G}_{[7,4]} = \begin{bmatrix} 1 & 0 & 0 & 0 & 1 & 1 & 1 \\ 0 & 1 & 0 & 0 & 1 & 1 & 0 \\ 0 & 0 & 1 & 0 & 1 & 0 & 1 \\ 0 & 0 & 0 & 1 & 0 & 1 & 1 \end{bmatrix}. \tag{7.21}
$$

By adding row 3 to row 1, then row 2 to row 1, then row 4 to row 3, row 3 to row 2, and finally, row 3 to row 1, we obtain the following sequence of equivalent generator matrices (note that leading and trailing zeros are not shown for better readability):

$$
\mathbf{G}_{[7,4]} \equiv \begin{bmatrix} 1 & 0 & 1 & 0 & 0 & 1 & \\ & 1 & 0 & 0 & 1 & 1 & \\ & & 1 & 0 & 1 & 0 & 1 \\ & & & 1 & 0 & 1 & 1 \end{bmatrix} \equiv \begin{bmatrix} 1 & 1 & 1 & 0 & 1 & & \\ & 1 & 0 & 0 & 1 & 1 & \\ & & 1 & 0 & 1 & 0 & 1 \\ & & & 1 & 0 & 1 & 1 \end{bmatrix}
$$

$$
\equiv \begin{bmatrix} 1 & 1 & 1 & 0 & 1 & & \\ & 1 & 0 & 0 & 1 & 1 & \\ & & 1 & 1 & 1 & 1 & \\ & & & 1 & 0 & 1 & 1 \end{bmatrix} \equiv \begin{bmatrix} 1 & 0 & 1 & 0 & 0 & 1 & \\ & 1 & 1 & 1 & & & \\ & & 1 & 1 & 1 & 1 & \\ & & & 1 & 0 & 1 & 1 \end{bmatrix}
$$

$$
\equiv \begin{bmatrix} 1 & 0 & 0 & 1 & 1 & & \\ & 1 & 1 & 1 & & & \\ & & 1 & 1 & 1 & 1 & \\ & & & 1 & 0 & 1 & 1 \end{bmatrix}. \tag{7.22}
$$

What we have generated is an equivalent minimum-span generator matrix (MSGM) [25] for the [7, 4] Hamming code, where the span of a matrix is defined as the sum of the spans of the rows, and the span of a row is defined as its length without the trailing and leading zeros. More precisely, the span of a vector $\mathbf{x}$ is defined as $\mathrm{Span}(\mathbf{x}) = R(\mathbf{x}) - L(\mathbf{x})$, where $R(\mathbf{x})$ is the index of the rightmost nonzero entry of $\mathbf{x}$, and, likewise, $L(\mathbf{x})$ is the index of the leftmost nonzero entry of $\mathbf{x}$. An MSGM is a generator matrix for $\mathcal{C}$ with minimum total span.

An MSGM can be obtained from an arbitrary generator matrix via the following simple algorithm:

Step 1. Find a pair of rows $\mathbf{x}^{(i)}$ and $\mathbf{x}^{(j)}$ in the generator matrix G such that $L(\mathbf{x}^{(i)}) = L(\mathbf{x}^{(j)})$ and $R(\mathbf{x}^{(i)}) \leq R(\mathbf{x}^{(j)})$, or $R(\mathbf{x}^{(i)}) = R(\mathbf{x}^{(j)})$ and $L(\mathbf{x}^{(i)}) \geq L(\mathbf{x}^{(j)})$.

Step 2. If Step 1 fails and no such pair can be found, go to Step 4.

Step 3.   Let $\mathbf{x}^{(i)} = \mathbf{x}^{(i)} + \mathbf{x}^{(j)}$; that is, replace the row $\mathbf{x}^{(i)}$ by the sum of the two rows. Go to Step 1.

Step 4.   Output G, which is now an MSGM.

We will now prove the following

THEOREM 7.7
The preceding algorithm always generates an MSGM.

PROOF
It is obvious from Step 3 in the algorithm that at each iteration the total span is reduced by one. The algorithm must therefore terminate in a finite number of steps, and it stops exactly when

$$L(\mathbf{x}^{(i)}) \neq L(\mathbf{x}^{(j)}), \qquad R(\mathbf{x}^{(i)}) \neq R(\mathbf{x}^{(j)}) \tag{7.23}$$

for all rows $\mathbf{x}^{(i)}, \mathbf{x}^{(j)}, \mathbf{x}^{(i)} \neq \mathbf{x}^{(j)}$, of $\mathbf{G}$.

We now need to show that no other generator matrix $\mathbf{G}'$ can have smaller span than the one just constructed, which we denote by $\mathbf{G}$. To this end, let $\mathbf{x}^{(1)}, \ldots, \mathbf{x}^{(k)}$ be the rows of $\mathbf{G}'$. Then, since $\mathbf{G}$ and $\mathbf{G}'$ are equivalent,

$$\mathbf{x}^{(j)\prime} = \sum_{\mathbf{x}^{(i)} \in \mathcal{I}_j} \mathbf{x}^{(i)}, \tag{7.24}$$

for every j, where $\mathcal{I}_j$ is some subset of the set of rows of $\mathbf{G}$. But due to (7.23) the sum in (7.24) can produce no cancellations at the endpoints of any member of $\mathcal{I}_j$:

$$\mathrm{Span}(\mathbf{x}^{(j)\prime}) = \bigcup_{\mathcal{I}_j} \mathrm{Span}(\mathbf{x}^{(i)}) \geq \max_i \mathrm{Span}(\mathbf{x}^{(i)}). \tag{7.25}$$

But the k-vectors $\mathbf{x}^{(j)\prime}$ must contain all $\mathbf{x}^{(i)}$'s, since otherwise $\mathbf{G}'$ has dimension $\leq k$ and cannot generate $\mathcal{C}$. Since every $\mathbf{x}^{(i)}$ is represented in at least one set $\mathcal{I}_j$, there exists an ordering of the indices i such that

$$\mathrm{Span}(\mathbf{x}^{(j)\prime}) \geq \mathrm{Span}(\mathbf{x}^{(i)}) \tag{7.26}$$

for every j, and hence also

$$\sum_{j=1}^{k} \mathrm{Span}(\mathbf{x}^{(j)\prime}) \geq \sum_{i=1}^{k} \mathrm{Span}(\mathbf{x}^{(i)}), \tag{7.27}$$

which proves the theorem.

We have actually proved that the matrix $\mathbf{G}$ has the smallest possible span for every row (up to row permutations); i.e., it has the smallest span set. From this we immediately deduce the following

COROLLARY 7.8

All MSGMs of a given code C have the same span set.

The significance of a MSGM lies in the fact that the dimension p_r of the past subcode P_r and the dimension f_r of the future subcode F_r, respectively, can be read off the generator matrix as explained by

THEOREM 7.9

Given an MSGM G,

$$p_r = |i : R(\mathbf{x}^{(i)}) \le r|; \tag{7.28}$$

i.e., p_r equals the number of rows in G for which the rightmost nonzero entry is at position r or before. Also,

$$f_r = |i : L(\mathbf{x}^{(i)}) \ge r + 1|; \tag{7.29}$$

i.e., f_r equals the number of rows in G for which the leftmost nonzero entry is at position $r + 1$ or later.

Proof

p_r is the dimension of the past subcode P_r, that is, the set of all codewords in C that merge with the zero state at position $r + 1$ or earlier. The rows of G are a basis for the code C; hence the rows of G that merge at time r or earlier, which are all independent, can be used as a basis for P_r. Due to (7.23), no other row, or linear combination of rows, can be in P_r, since then a codeword would have nonzero entries x_j, for $j > r$, and therefore every codeword in P_r is a linear combination of said rows. There are exactly $|i : R(x^{(i)}) \le r|$ qualifying rows. This together with Corollary 7.8 proves (7.28).

Analogously, the rows of G that start at position $r + 1$ or later generate the subcode F_r. Since there are $|i : L(x^{(i)}) \ge r + 1|$ independent rows with that property, this proves (7.29).

We summarize that the PC trellis is optimal in that it simultaneously reduces the number of states and edges in its trellis representation of a linear block code C and, hence, would logically be the trellis of choice. Although the PC trellis is optimal for a given code, bit position permutations in a code, which generate equivalent codes, may bring further benefits. The problem of minimizing trellis complexity allowing such bit permutations is addressed in [27–29].

We now use Theorem 7.9 to read the values of p_r and f_r off the MSGM $G_{[7,4]}$, given in (7.22). This produces the following table:

r	1	2	3	4	5	6	7	8
p_r	0	0	0	0	1	2	3	4
f_r	4	3	2	1	0	0	0	0

From this table we calculate $|\mathcal{S}_r| = 2^{k-f_r-p_r}$ and $|\mathcal{B}_{r,r+1}| = 2^{k-f_{r+1}-p_r}$, given by

r	1	2	3	4	5	6	7	8
$|\mathcal{S}_r|$	1	2	4	8	8	4	2	1
$|\mathcal{B}_{r,r+1}|$	2	4	8	16	8	4	2	—

These values correspond exactly to those in Figure 7.1, where we have constructed the trellis explicitly.

7.5 CONSTRUCTION OF THE PC TRELLIS

The construction of the PC trellis in Section 7.3 was relatively simple because the code was systematic and we did not have to deal with a large code. In this section we discuss a general method to construct the PC trellis from an MSGM of a block code. This method was presented by McEliece in [25]. We will use the fact that the rows of G form bases for P_r and F_r according to Theorems 7.9 and 7.6; that is, $|\mathcal{S}_r| = 2^{k-p_{r-1}-f_{r-1}}$. Let us start with an example and consider the MSGM G for the extended $[8, 4]$ Hamming code, given by

$$G_{[8,4]} = \begin{bmatrix} 1 & 1 & 1 & 1 & 0 & 0 & 0 & 0 \\ 0 & 1 & 0 & 1 & 1 & 0 & 1 & 0 \\ 0 & 0 & 1 & 1 & 1 & 1 & 0 & 0 \\ 0 & 0 & 0 & 0 & 1 & 1 & 1 & 1 \end{bmatrix}. \tag{7.30}$$

At $r = 1$ we have a single starting state, $|\mathcal{S}_r| = 1|$. By inspection we see that from $r = 1$ to $r = 3$ the first row $\mathbf{x}^{(1)} = (11110000)$ is active. By that we mean that any codeword that contains $\mathbf{x}^{(1)}$ will be affected by it in positions $r = 1$ to $r = 3 + 1$. Likewise, the second row $\mathbf{x}^{(2)}$ is active for $r \in [2, 6]$, $\mathbf{x}^{(3)}$ is active for $r \in [3, 5]$, and $\mathbf{x}^{(4)}$ is active for $r \in [5, 7]$. The basic idea of the trellis construction is that each row in the MSGM is needed as a basis only where it is active, and each active row doubles the number of states. We will now formalize this more precisely. Let w_r be a vector with dimension σ_r whose ith entry is the ith active row. These vectors are given for $G_{[8,4]}$ above as

w_0	w_1	w_2	w_3	w_4	w_5	w_6	w_7	w_8
—	$[w^{(1)}]$	$\begin{bmatrix} w^{(1)} \\ w^{(2)} \end{bmatrix}$	$\begin{bmatrix} w^{(1)} \\ w^{(2)} \\ w^{(3)} \end{bmatrix}$	$\begin{bmatrix} w^{(2)} \\ w^{(3)} \end{bmatrix}$	$\begin{bmatrix} w^{(2)} \\ w^{(3)} \\ w^{(4)} \end{bmatrix}$	$\begin{bmatrix} w^{(2)} \\ w^{(4)} \end{bmatrix}$	$[w^{(4)}]$	—

Further, let $\tilde{w}_r$ be a k-vector with entries from w_r in the appropriate places and padded with zeros; for example, $\tilde{w}_6 = (0, w_{(2)}, 0, w_{(4)})$. We now let the states $\mathcal{S}_{r+1}$ be the set $\{w_r\}$ with $w_r^{(i)} \in \{0, 1\}$. This gives the states of the trellis in Figure 7.4. Two states in this trellis are connected if either $w_r \in w_{r+1}$ or $w_{r+1} \in w_r$; for example, $w_2 = (01)$ and $w_3 = (011)$ are connected since $w_2 = (w^{(1)} = 0, w^{(2)} = 1)^T$ is contained in $w_3 = (w^{(1)} = 0, w^{(2)} = 1, w^{(3)} = 1)$. The trellis branch labels into state w_{r+1} are given by

$$x_r = \left(w_r \bigcup \widetilde{w_{r+1}} \right) \cdot \mathbf{g}_r, \tag{7.31}$$

where $\mathbf{g}_r$ is the rth column in the generator matrix. As an example, x_4 on the transition $(111)^T \rightarrow (11)^T$ is given by $(w^{(1)}, \widetilde{w^{(2)}}, w^{(3)}) \cdot \mathbf{g}_r = (1110) \cdot (1110) = 1$. Since the state vector w_r indicates which active rows are present in the codewords that pass through w_r, it should be easy to see that equation (7.31) simply equals the symbol that these active rows generate at time r.

If we combine pairs of branches into single branches with two symbols as branch labels, the resulting new trellis has a maximum number of states of only four, half that of the original PC trellis. This new trellis is shown in Figure 7.5. We construct this trellis again in Section 7.7 in our

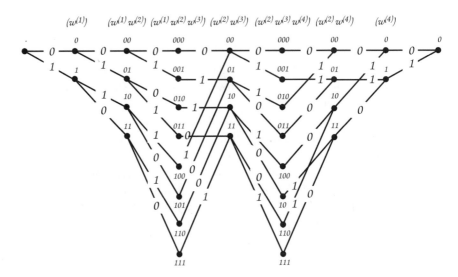

Figure 7.4 PC trellis for the [8, 4] extended Hamming code, constructed in the systematic way. States are labeled as row vectors.

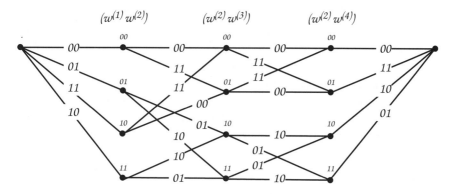

Figure 7.5 Contracted PC trellis for the [8, 4] extended Hamming code, obtained by taking pairs of symbols as branches.

discussion of Reed-Muller codes. It becomes apparent that there are many ways of representing a block code, and it is not always straightforward to determine the best method. It seems to us, however, that the PC trellis has the most claim to being the minimal complexity trellis, despite the situation just described.

7.6   THE SQUARING CONSTRUCTION AND THE TRELLIS OF LATTICES

In this section we take a reversed approach to that of the previous sections; that is, we construct lattices and block codes by constructing their trellises. Lattices were used in Chapter 3 as signal sets for trellis codes, and we now learn that they too can be described by trellises. To this end we introduce a general building block, the squaring construction, that has found application mainly in the construction of lattices [31].

Let us assume that some set S of signals is the union of M disjoint subsets T_i. This is a partition of S into M subsets, and we denote this partition by S/T. Furthermore, let $d(S)$ be the minimum Euclidean distance between any two elements of S, that is, $d(S) = \min_{s_1,s_2 \in S(s_1 \neq s_2)} |s_1 - s_2|^2$, and let $d(T_j)$ be the minimum distance between any two elements of the subset T_j. Define $d(T) = \min_j T_j$ as the minimum distance between elements in any of the subsets. As an example consider the lattice partition $\Lambda = \Lambda' + [\Lambda/\Lambda']$ from Chapter 3, where the subsets are the cosets of the sublattice Λ'.

We now define the *squaring construction* [30] as the set of signals U, given by

DEFINITION 7.2

The union set U, also denoted by $|S/T|^2$, resulting from the squaring construction of the partition S/T is the set of all pairs (s_1, s_2), such that $s_1, s_2 \in T_j$; that is, s_1 and s_2 are in the same subset T_j.

The squaring construction can conveniently be depicted by a trellis diagram (Figure 7.6). This trellis has M states, corresponding to the M subsets T_j of S. This is logical, since we need to remember the subset of s_1 in order to restrict the choice of s_2 to the same subset.

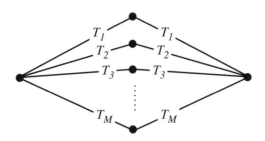

Figure 7.6 Trellis diagram of the squaring construction.

Continuing our lattice example, let us apply the squaring construction to the lattice partition $Z^2/RZ^2 = Z^2/D_2$ (compare Figure 3.14), where T_1 is the lattice RZ^2 and T_2 is its coset $RZ^2 + (0, 1)$. This is illustrated in Figure 7.7, and we see that this construction yields D_4, the Schläfli lattice, whose points have an integer coordinate sum. (The points in RZ^2 have even coordinate sums, and the points in its coset $RZ^2 + (0, 1)$ have odd coordinate sums.)

Note that in applying the squaring construction to Z^2/D_2 we have obtained another lattice, D_4; that is, any two points $d_1, d_2 \in D_4$ can be added as four-dimensional vectors to produce $d_3 = d_1 + d_2 \in D_4$, another

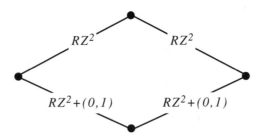

Figure 7.7 Using the squaring construction to generate D_4 from the partition Z^2/D_2.

point in the lattice. That this is so can be seen from Figure 7.7. This property results of course from the fact that RZ_2 is a subgroup of Z_2 under vector addition (i.e., a sublattice), inducing the group partition Z_2/RZ_2. Our lattice examples have therefore more algebraic structure than strictly needed for the squaring construction.

The squaring construction is an important tool for the following.

LEMMA 7.10

Given the distances $d(T)$ and $d(S)$ in the partition S/T, the minimum distance $d(U)$ of U obeys

$$d(U) = \min[d(T), 2d(S)]. \tag{7.32}$$

Proof

There are two cases we need to examine:

i   If the two elements u_1, u_2 achieving $d(U)$ have their first components, $t_1, t_2 \in T_j$, $t_1 \neq t_2$, in the same subset, that is, $u_1 = (t_1, s_1)$, $u_2 = (t_2, s_2)$, where $s_1 = s_2$, then their distance $|u_1 - u_2|^2 = d(T)$, and, consequently, $d(U) = d(T)$.

ii If their first components t_1, t_2 lie in different subsets T_j and T_i, their second components s_1, s_2 must lie in the same two respective subsets by virtue of the squaring construction, i.e., $s_1 \in T_j$ and $s_2 \in T_i$, since $T_j \neq T_i, \Rightarrow s_1 \neq s_2$. Likewise, $t_1 \neq t_2$, and, since t_1, t_2, s_1, and s_2 can be chosen such that $d(t_1, t_2) = d(S)$ and $d(s_1, s_2) = d(S)$, we conclude that $d(U) = 2d(S)$ in this case. This proves the lemma.

We now extend the squaring construction to two levels. Let $S/T/V$ be a two-level partition chain; that is, each subset T_j is made up of P disjoint subsets V_{ji}. Let $T_j \times T_j = |T_j|^2$ be the set of all elements (t_1, t_2), $t_1, t_2 \in T_j$. Then each $|T_j|^2$ can be represented by the trellis in Figure 7.8; that is, each element $t_1 \in V_{ji}$ can be combined with a second element t_2 from any $V_{ji} \in T_j$, and we have a two-stage trellis with P states representing $|T_j|^2$ in terms of its subsets V_{ji}. Since each $|T_j|^2$ can be broken down into a trellis like the one in Figure 7.8, we are led to the two-level representation of the set U in Figure 7.9. It should now be straightforward to see that Figures 7.6 and 7.9 show the same object with different degrees of detail.

There is another way of looking at Figure 7.8. The paths to the first node at the second stage (solid paths in the figure) are the set obtained by the squaring construction of the partition T_j/V_j, denoted by W_j. Obviously, W_j is a subset of $|T_j|^2$. In fact $|T_j|^2$ is the union of P sets W_{ji}, each of

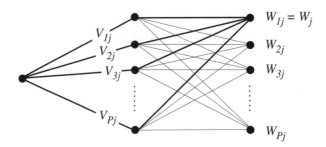

Figure 7.8 Trellis representation of the set $|T_j|^2 = T_j \times T_j$, using the second-level partition T/V.

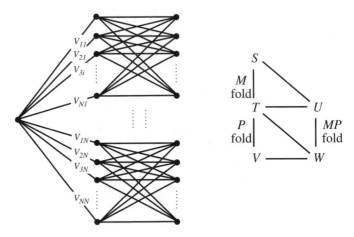

Figure 7.9 Representation of the original squaring construction refined to the two-level partition $S/T/V$ and schematic representation of the induced partition U/W.

which, except for W_j, is obtained from a squaring-type construction similar to the squaring construction but with the indices of the second component sets permuted cyclically. This is called a *twisted squaring construction*.

This then induces an MP partition of the set U, denoted by U/W, as shown in Figure 7.9; i.e., the set U is partitioned into MP disjoint sets W_{ji}. The right side of Figure 7.9 is a schematic diagram of this construction. The partition chain $S/T/V$ induces a partition U/W via the squaring construction, whereby U is obtained from the squaring construction $|S/T|^2$ and W from $|T/V|^2$. Since the partition S/T is P-fold and the partition T/V is M-fold, the resulting partition U/W is MP-fold.

Well, we may now apply the squaring construction to the new partition U/W and obtain a *two-level squaring construction* (Figure 7.10). This is done by concatenating back to back two trellis sections of the type in

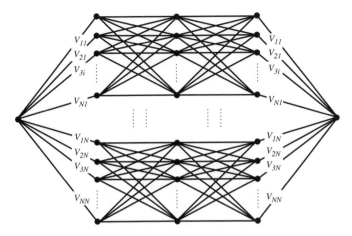

Figure 7.10 Two-level squaring construction for the partition $S/T/V$.

Figure 7.9, just as in the one-level squaring construction. We denote the two-level squaring construction by $|S/T/V|^4$, which is identical to the one-level squaring construction $|U/W|^2$. Using Lemma 7.10 twice we obtain

$$d(|S/T/V|^4) = \min[d(W), 2d(U)]$$
$$= \min[d(V), 2d(T), 4d(S)]. \tag{7.33}$$

As an example of the two-level squaring construction, let us start with the binary lattice partition chain of the integer lattice, $Z^2/RZ^2/2Z^2$. The resulting lattice[2] has eight dimensions and its four-state trellis diagram is shown in Figure 7.11. Its minimum squared Euclidean distance between lattice points is $d(|Z^2/RZ^2/2Z^2|^4) = 4$. This is the famous Gosset lattice E_8 [31].

This game of two-level squaring constructions may now be continued with the newly constructed lattice; that is, we can recursively construct the lattices $\Lambda(n) = |\Lambda(n-1)/R\Lambda(n-1)|^2 = |\Lambda(n-2)/R\Lambda(n-2)/2\Lambda(n-2)|^4$ starting with $D_4 = \Lambda(1) = |Z^2/RZ^2|^2$ and $E_8 = \Lambda(2) = |Z^2/RZ^2/2Z^2|^4$. The sequence of these lattices is known as the sequence of *Barnes-Wall* lattices $\Lambda_N = \Lambda(n)$ of dimension $N = 2^{n+1}$ and minimum squared Euclidean distance $2^n = N/2$ [30, 31].

The minimum squared distance results from the squaring construction; that is,

$$d(\Lambda_N) = \min\left(d(R\Lambda_{N/2}), 2d(\Lambda_{N/2})\right)$$
$$= \min\left(d(2\Lambda_{N/4}), 2d(R\Lambda_{N/4}), 4d(\Lambda_{N/4})\right).$$

---

[2] The fact that the resulting union set is also a lattice is argued analogously to the case of D_4.

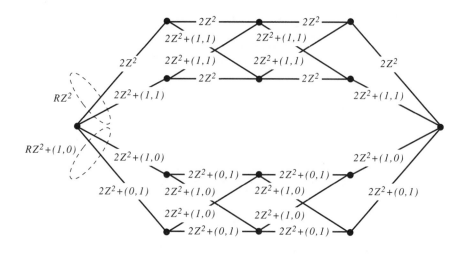

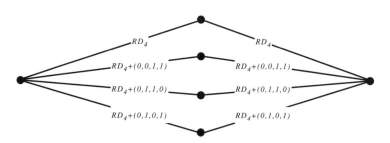

Figure 7.11 Gosset lattice E_8 obtained via the two-level squaring construction. The subsets are expressed as cosets $Z_2 + \mathbf{c}$, where $\mathbf{c}$ is the coset representative (see also Section 3.5). The lower figure shows E_8 as the single-level squaring construction from the partition $U/V = D_4/RD_4$.

The distance sequence now follows by induction, starting with $d(\Lambda_2 = Z^2) = 1$ and $d(\Lambda_4 = D_4) = 2$ and using the general fact that $d(R\Lambda) = 2d(\Lambda)$; that is, $d(\Lambda_N) = 2d(\Lambda_{N/2}) = 4d(\Lambda_{N/4})$, and hence $d(\Lambda_N) = N/2$.

Note that, as can be seen from the squaring construction, Λ_N has the same minimum distance as $R\Lambda_{N/2} \times R\Lambda_{N/2}$, namely $N/2$, but has $2^{N/4}$ times as many lattice points ($2^{N/4}$ is also the number of cosets in the $\Lambda_{N/2}/R\Lambda_{N/2}$ partition). The asymptotic coding gain of Λ_N, given by $\gamma(\Lambda_N) = d(\Lambda_N)/V(\Lambda_N)^{2/N}$ (see equation (3.9)), is therefore $2^{1/2}$ times that of $R\Lambda_{N/2} \times R\Lambda_{N/2}$, which is also the coding gain of $\Lambda_{N/2}$. Starting with $\gamma(Z^2) = 1$, this leads to an asymptotic coding gain of the

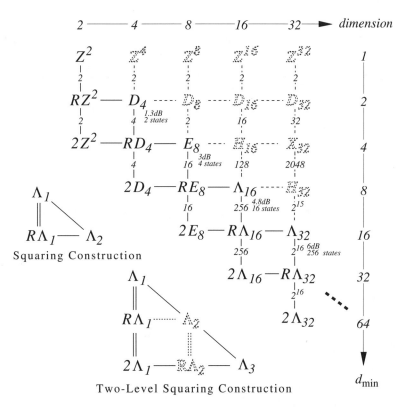

Figure 7.12 Schematic representation of the relationships of the infinite sequence of Barnes-Walls and related lattices. The small numbers indicate the order of the lattice partition $|\Lambda / \Lambda'|$. The asymptotic coding gain and the number of trellis states are also given for the main sequence of lattices.

Barnes-Walls lattices given by $\gamma = 2^{n/2} = N/2$, which increases without bound. The relationships and construction of the infinite sequence of Barnes-Walls lattices is schematically represented in Figure 7.12, which also includes the relevant parent lattices.

Since the partitions $\Lambda_N / R\Lambda_N$ and $R\Lambda_N / 2\Lambda_N$ are both of the order $2^{N/2}$, the number of states in the trellis diagram for Λ_{4N} resulting from the two-level squaring construction is 2^N, and therefore the number of states in the trellis of the Barnes-Walls lattices Λ_N is given by $2^{N/4} = 2^{2^{n-1}}$, which grows exponentially with the dimension N. Thus, we have 2 states for D_4, 4 states for E_8, 16 states for Λ_{16}, and then 256, 65,536, and 2^{64} states for Λ_{32}, Λ_{64}, and Λ_{128}, respectively, and the number of states, or the complexity of the lattice, also increases without bound.

If we want to use E_8 as a code for a telephone line modem, we would choose quadrature modulation and transmit four two-dimensional signals to make up one eight-dimensional codeword or lattice point. A typical baud rate over telephone channels is 2400 symbols/s (baud). To build a modem that transmits 9600 bits/s, we require 16 signal points every two dimensions, and the total number of signal points from E_8 is $2^{16} = 65,536$. To transmit 14 400 bits/s we already need 2^{24} signal points, approximately 17 million. It becomes evident that efficient decoding algorithms are needed since exhaustive look-up tables clearly become infeasible. The trellis structure of the lattices provides an excellent way of breaking down the complexity of decoding, and we will see in Section 7.8 that a decoder for the E_8 lattice becomes rather simple indeed. Further aspects of implementation of lattice modems are discussed by Lang and Longstaff [32] and by Conway and Sloane [33].

7.7   THE CONSTRUCTION OF REED-MULLER CODES

In this section we start with the two-level partition chain $(2, 2)/(2, 1)/(2, 0)$ of length-2 binary codes. $(2, 2)$ is the binary code over GF(2) consisting of all four length-2 binary codewords. $(2, 1)$ is its subcode consisting of the codewords $\{(0, 0), (1, 1)\}$, and $(2, 0)$ is the trivial single codeword $(0, 0)$. Since $(2, 1)$ is a subgroup of $(2, 2)$ under vector addition over GF(2), $(2, 2)/(2, 1)$ is a true partition, and the same holds for $(2, 1)/(2, 0)$. We now define [30] the Reed-Muller codes $RM(r, n)$ with parameters r, n recursively as

$$RM(r, n) = |RM(r, n - 1)/RM(r - 1, n - 1)|^2, \qquad (7.34)$$

and start the recursion with $RM(-1, 1) = (2, 0)$, $RM(0, 1) = (2, 1)$, and $RM(1, 1) = (2, 2)$. The first two codes constructed are illustrated in Figure 7.13. The code $RM(1, 2) = |(2, 2)/(2, 1)|^2$ is a $(4, 3)$ single parity-check code, and the code $RM(0, 2) = |(2, 1)/(2, 0)|^2$ is the $(4, 1)$ repetition code.

Figure 7.14 shows the construction tableau for the Reed-Muller codes. Every column is completed at the top with $RM(n, n)$, the binary $(2^n, 2^n)$ code consisting of all length-2^n binary vectors, and at the bottom with $RM(-1, n)$, the binary code consisting of the single codeword $(0, \ldots, 0)$ of length 2^n. The partition orders are also indicated in the figure.

We now need to determine the code parameters of the $RM(r, n)$ codes. Their length $N = 2^n$, and, from Lemma 7.10, we establish recursively the minimum distance $d_{min} = 2^{n-r}$. The code rate is found as follows. The

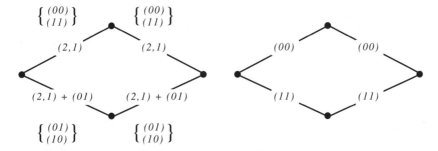

Figure 7.13 The construction of RM(1, 2) and RM(0, 2) via the squaring construction from the binary codes (2, 2), (2, 1), and (2, 0).

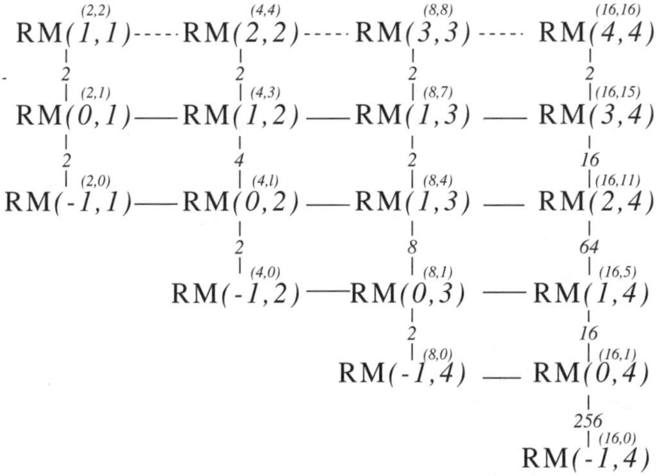

$$
\begin{array}{cccc}
\overset{(2,2)}{\text{RM}(1,1)} & \!\!\!----\!\!\! & \overset{(4,4)}{\text{RM}(2,2)} & \!\!\!----\!\!\! & \overset{(8,8)}{\text{RM}(3,3)} & \!\!\!----\!\!\! & \overset{(16,16)}{\text{RM}(4,4)}
\end{array}
$$

$$
\begin{array}{c}
2 \quad\qquad 2 \qquad\qquad 2 \qquad\qquad 2 \\
\overset{(2,1)}{\text{RM}(0,1)}\!\!-\!\!\overset{(4,3)}{\text{RM}(1,2)}\!\!-\!\!\overset{(8,7)}{\text{RM}(1,3)}\!\!-\!\!\overset{(16,15)}{\text{RM}(3,4)}
\end{array}
$$

$$
\begin{array}{c}
2 \quad\qquad 4 \qquad\qquad 2 \qquad\qquad 16 \\
\overset{(2,0)}{\text{RM}(-1,1)}\!\!-\!\!\overset{(4,1)}{\text{RM}(0,2)}\!\!-\!\!\overset{(8,4)}{\text{RM}(1,3)}\!\!-\!\!\overset{(16,11)}{\text{RM}(2,4)}
\end{array}
$$

$$
\begin{array}{c}
2 \qquad\qquad 8 \qquad\qquad 64 \\
\overset{(4,0)}{\text{RM}(-1,2)}\!\!-\!\!\overset{(8,1)}{\text{RM}(0,3)}\!\!-\!\!\overset{(16,5)}{\text{RM}(1,4)}
\end{array}
$$

$$
\begin{array}{c}
2 \qquad\qquad 16 \\
\overset{(8,0)}{\text{RM}(-1,4)}\!\!-\!\!\overset{(16,1)}{\text{RM}(0,4)}
\end{array}
$$

$$
\begin{array}{c}
256 \\
\overset{(16,0)}{\text{RM}(-1,4)}
\end{array}
$$

Figure 7.14 Diagram for the construction of the Reed-Muller codes via the squaring and two-level squaring constructions.

partition order of $|\text{RM}(r, n)/\text{RM}(r - 1, n)|$, denoted by $m(r, n)$, follows from the squaring construction and obeys the recursion

$$m(r, n) = m(r, n - 1)m(r - 1, n - 1). \qquad (7.35)$$

Starting this recursion with $m(1, 1) = 2$ and $m(0, 1) = 2$ leads to the partition numbers in Figure 7.14. Converting to logarithms, that is, $M(r, n) = \log_2(m(r, n))$, (7.35) becomes

$$M(r, n) = M(r, n - 1) + M(r - 1, n - 1), \qquad (7.36)$$

which, after initialization with $M(1, 1) = 1$ and $M(0, 1) = 1$ generates Pascal's triangle, whose numbers can also be found via the combinatorial generating function $(1 + x)^n$. The information rates of the $RM(r, n)$ codes are now found easily. The rate of $RM(r, n)$ is given by the rate of $RM(r - 1, n)$ plus $M(r, n)$. This generates the code rates indicated in Figure 7.14.

Among the general class of Reed-Muller codes we find the following special classes:

- $RM(n - 1, n)$ are the single parity-check codes of length 2^n.
- $RM(n - 2, n)$ is the family of extended Hamming codes of length 2^n with minimum distance $d_{\min} = 4$ [30, 31].
- $RM(1, n)$ are the first-order Reed-Muller codes of length $N = 2^n$, rate $R = n + 1$, and minimum distance $d_{\min} = N/2$ [31].
- $RM(0, n)$ are the repetition codes of length 2^n.

Figure 7.15 shows the four-section trellis diagrams of $RM(2, 4)$, a $[16, 11]$ code, and of $RM(1, 4)$, a $[16, 5]$ code.

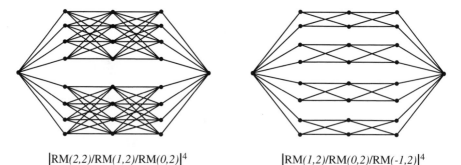

$|RM(2,2)/RM(1,2)/RM(0,2)|^4$ $|RM(1,2)/RM(0,2)/RM(-1,2)|^4$

Figure 7.15 The trellis diagrams of $RM(2, 4)$ and $RM(1, 4)$, constructed via the two-level squaring construction.

7.8   A DECODING EXAMPLE

We discussed trellis decoding procedures at length in Chapter 6, and all of those methods are applicable to trellis decoding of block codes. In particular, the trellises of the codes and lattices constructed in the preceding sections lend themselves to efficient decoder implementations due to their regular structure. The construction also resulted in efficient trellises in terms of their numbers of states. The extended Hamming codes constructed in Section 7.7 have $N/2$ states, while the PC trellises of the original and the extended Hamming codes, constructed according to Section 7.3, have

$2^{n-k} = N$ states, a savings of a factor of 2, obtained by taking pairs of coded bits as branch symbols.

Often the trellis decoding operation can be mapped into an efficient decoder. Let us consider the extended Hamming code $RM(1, 3)$, whose trellis is shown in Figure 7.16. Note that the trellis is the same as that of the Gosset lattice E_8 in Figure 7.11. The decoder for both is therefore essentially the same. Let us then assume that, as is the usual practice in telecommunications, that zeros are mapped into -1's and 1's are retained; that is, we map the output signals $\{0, 1\}$ of the code into a BPSK signal set and we obtain the modified codewords $\mathbf{x}'$ from the original $\mathbf{x}$. With this the decoder now operates as follows. For each of the signals y_i from the received signal vector $\mathbf{y} = (y_1, \ldots, y_8)$, two metrics need to be computed: one is $(y_i - 1)^2$, and the other $(y_i + 1)^2$ or, equivalently, $m_0 = y_i$ and $m_1 = -y_i$. We see that m_0 and m_1 are negatives of each other, and only one must be calculated. Since the all-one vector $\mathbf{x} = (1, \ldots, 1)$ is a codeword, the negative of every modified codeword $\mathbf{x}'$ is itself a codeword, as can be seen by applying the linearity of the code. The sign of the metric sum can now be used to identify the sign of the codeword.

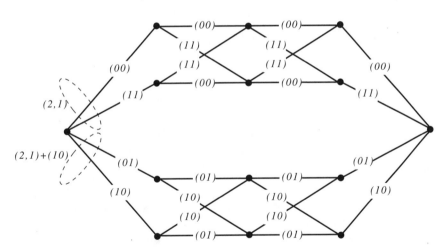

Figure 7.16 Trellis of the [8, 4] extended Hamming code. This is also the trellis of the Gosset lattice from Figure 7.11 as can be seen by comparing Figures 7.12 and 7.14.

The operation of the decoder is illustrated in Figure 7.17, where the operator $\mathcal{O}$ produces the sum and difference of the inputs; that is, the outputs of the first operator box are $y_1 + y_2$ and $y_1 - y_2$. The first two stages in the decoder decode the first and the second trellis sections, and the sign

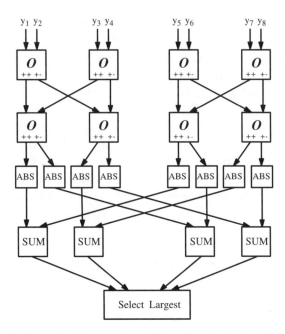

Figure 7.17 Schematic diagram of an efficient decoder for the [8, 4] extended
Hamming code. The operator O produces the sum $y_i + y_j$ and the
difference $y_i - y_j$ of the inputs, ABS forms the absolute value and
SUM adds $y_i + y_j$.

indicates whether the codeword or its complement are to be selected. After
the absolute magnitude operation, the two sections are combined. This
is done by the last section of the decoder. Finally, the selection decides
through which of the four central states the best codeword leads. Since
the components have already been decoded, this determines the codeword
with the best total metric.

In order to use the above decoder to decode E_8, one additional initial
step needs to be added. For every y_i, the closest even integer I_e and the
closest odd integer I_o need to be found first, and the metrics become $m_0 = (I_0 - y_i)^2$ and $m_1 = (I_e - y_i)^2$. These integers need to be stored to
reconstruct the exact lattice point. The remainder of the decoding operation
is identical to that of the extended Hamming code.

7.9   TREE-BASED SOFT-DECISION DECODING
OF BLOCK CODES

Traditionally block codes have been decoded algebraically, via error trap-
ping or threshold decoding [19]. This required hard decisions on the

received channel symbols, and hence block codes are often known also as *error-correcting codes*. Soft-decision decoding of block codes, which provides additional gain, is very difficult and cumbersome using an algebraic approach. Additionally, maximum-likelihood (ML) performance often cannot be accomplished and performance falls short of optimum.

In the last sections we have explored in some detail the trellis structure of block codes, and using this trellis representation we may use any of the decoding algorithms from Chapter 6 to decode block codes. In particular, the minimal trellis of Section 7.4 and the constructed trellises of Sections 7.6 and 7.7 are well suited for the application of an ML search algorithm, namely the Viterbi algorithm.

In this section we apply the M-algorithm (Section 6.4) to the problem of decoding block codes. For the M-algorithm we do not need the code trellis, a simpler representation as a tree will be sufficient. We will see that with very little effort this suboptimal decoding procedure performs nearly as well as optimal ML decoding, without requiring the minimal trellis of the code.

Consider again the [7, 4] Hamming code from Section 7.3. In its systematic form (7.2) this code has the generator matrix

$$G_{[7,4]} = \begin{bmatrix} 1 & & & & 0 & 1 & 1 \\ & 1 & & & 1 & 0 & 0 \\ & & 1 & & 0 & 1 & 0 \\ & & & 1 & 0 & 0 & 1 \end{bmatrix}. \tag{7.37}$$

We can represent this code by the tree in Figure 7.18. This tree is identical to the trellis of Figure 7.1 without the mergers. Each terminal node in the code tree represents a distinct path through the trellis of Figure 7.1 from $s_1^{(0)} \rightarrow s_{n+1}^{(0)}$. This tree has a growing section of length $k = 4$ and a tail section of length $n - k = 3$. Its number of terminal nodes equals the total number of codewords, $2^k = 16$. The growing section is identical for all $[n, k]$ block codes, since the branch labels are simply the information bits. We call this part the information portion of the code tree. The branch labels of the tail section can be found easily from (7.37), that is, by simply reencoding the information branch labels of the corresponding section in the information portion of the code tree, which we call the parity portion of the code tree. This straightforward generation of the code tree stands in marked contrast to the elaborate procedure required for finding the minimal code trellis so important for Viterbi decoding.

It is of course not feasible to search the entire code tree for large k, and we wonder whether this more simplistic view of a block code is affecting the performance of the decoder. It turns out that this is not the case, at

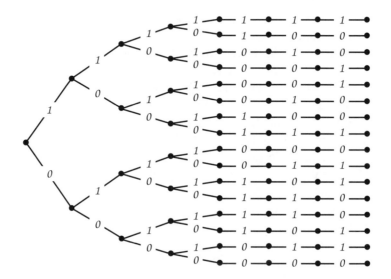

Figure 7.18 Code tree of the [7, 4] Hamming code whose trellis is shown in Figure 7.1.

least for short codes. Recall that in the M-algorithm we retain only the M paths with the highest current metric at each level in the tree. Note also that up to depth k in the code tree no information about the code is needed since the branch labels in the growing section of the code tree are simply the information bits. Beyond level k the decoder requires a copy of the encoder in order to produce the labels on the parity portion of the code tree.

Simulations of several block codes using the M-algorithm have been performed by Jelicic et al. [34], and we present some of their results in Figures 7.19–7.21. In all simulations BPSK signaling is assumed (Chapter 2); that is, the coded bits $(1, 0)$ map into the channel symbols $(+\sqrt{RE_b}, -\sqrt{RE_b})$, where $R = k/n$ and E_b is the received energy per bit.

Figure 7.19 presents simulation results of the [7, 4] Hamming code using the M-algorithm with $M = 4$. Also shown is the performance of hard-decision decoding using algebraic decoding, as well as the performance of optimal ML decoding using the Viterbi algorithm, which requires eight states. The M-algorithm comes to within a fraction of a dB of optimal performance.

Figure 7.20 likewise shows simulation results for the two (hard-decision) error-correcting [15, 7] BCH codes. Again, the M-algorithm achieves near-optimal performance with $M = 16$, whereas ML decoding requires $2^7 = 128$ states.

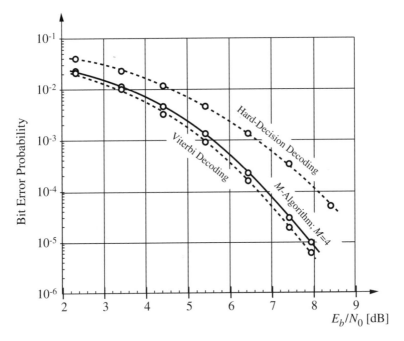

Figure 7.19 Simulation results using the M-algorithm with the [7, 4] Hamming code. Also shown are the performance curves of algebraic hard-decision decoding and optimal soft-decision decoding using the code trellis and the Viterbi algorithm.

Figure 7.21 finally presents simulation results for the [23, 12] Golay code. Near-optimal performance is achieved with $M = 64$, and optimal performance requires a Viterbi decoder with $2^{11} = 2048$ states.[3]

Since the complexity of retaining a path and that of maintaining a state in the ML decoder are roughly comparable, we observe that a dramatic savings in complexity can be made by using the M-algorithm versus Viterbi decoding in the general case. The cost paid for this is an almost negligible loss in performance.

Attaining ML performance with the M-algorithm would require storage of all paths, and hence large complexity. However, our observations point toward a hypothesis that most of a code's coding gain can be achieved with a small fraction of that complexity. It is that last fraction of a dB that

---

[3] Just like the [8, 4] extended Hamming code had a four-state contracted trellis (Figures (7.4) and (7.5)), the [24, 12] extended Golay code has a three-section 64-state trellis diagram, whose branch labels are 8-bit sequences [30]. For a highly structured code like the Golay code, maximum likelihood decoding may therefore not be an infeasible exercise in complexity. Nevertheless the reader may want to keep in mind that the low-complexity-tree approach with the M-algorithm can be used for any linear block code without reference to the internal structure of the code.

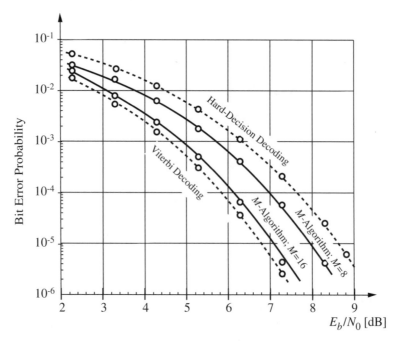

Figure 7.20 Simulation results using the M-algorithm to decode the [15, 7] BCH
        code.

costs most. Similar observations about the efficiency of the M-algorithm
have been made in the application of the M-algorithm to the problem of
multiuser interference cancellation [36].

In the remainder of this section we address the question of how large
M has to be in order to achieve near-optimal performance. In the case of
decoding block codes this question has a relatively simple answer. According to (6.23), the block error probability $\overline{P_B}$ of the M-algorithm can
be bounded by

$$\overline{P_B} \leq \overline{P_{B,\mathrm{ML}}} + \Pr(\mathrm{CPL}). \tag{7.38}$$

We now define the performance of the M-algorithm to be *near optimal*
if the probability of path loss is at most as large as the block error probability
of ML decoding. We then need to find M such that $\Pr(\mathrm{CPL}) < \overline{P_{B,\mathrm{ML}}}$.
In this case $\overline{P_B}$ is dominated by $\overline{P_{B,\mathrm{ML}}}$, and the performance of the M-
algorithm is close to that of an ML decoder. We need to find $\min(d_l^2)$ to
ensure (6.22), and the technique in Appendix 6.A can be used. However,
we find that due to the special structure of the block code tree, an intuitive
geometric approach is much more illustrative.

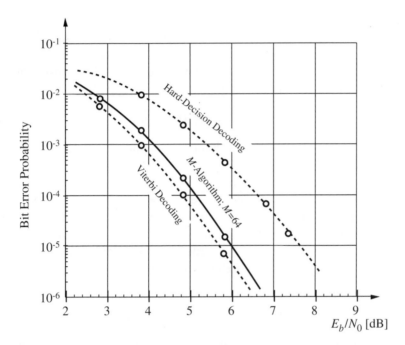

Figure 7.21 Simulation results of the M-algorithm used to decode the $[23, 12]$ Golay code.

Following [34, 35], we break Pr(CPL) into components

$$\text{Pr(CPL)} = \text{Pr}_{\text{info(CPL)}} + (1 - \text{Pr}_{\text{info}}(\text{CPL}))\,\text{Pr}_{\text{parity}}(\text{CPL}), \qquad (7.39)$$

where $\text{Pr}_{\text{info}}(\text{CPL})$ is the probability that we lose the correct path in the information section of the code tree, and $\text{Pr}_{\text{parity}}(\text{CPL})$ is the probability that we loose the correct path in the parity section of the tree.

The signal points associated with the partial codewords in the information section all lie on the vertices of an r-dimensional hypercube ($r \leq k$). Each coordinate of these corner points is $\pm\sqrt{RE_b}$. The minimum distance between two signal points in this portion of the code tree is therefore $2\sqrt{RE_b}$.

The problem of finding $\text{Pr}_{\text{info}}(\text{CPL})$ is completely symmetrical, and we may look at losing the correct path from any one of the corner points of the hypercube; that is, we may choose any codeword as the correct one, and we henceforth have the all-zero codeword do this honor.

If $M = 1$, $\text{Pr}_{\text{info}}(\text{CPL})$ is approximated by the probability of mistaking two nearest neighbors, $\text{Pr}_{\text{info}}(\text{CPL}) \approx Q\left(\sqrt{2RE_b/N_0}\right)$. For $M = 2$, $\text{Pr}_{\text{info}}(\text{CPL})$ is the probability that the noise moves the signal point away from two adjacent signal points; that is, the noise moves the signal point outside two adjacent decision regions. It can be seen quite easily [35] that this probability is given by the probability that the noise carries the transmitted signal point (00) into the shaded region in Figure 7.22. Consideration of all permissible sets of retained path leads to the upperbound

$$\Pr(\text{CPL}) \leq \kappa Q\left(\sqrt{4RE_b/N_0}\right), \tag{7.40}$$

where $\kappa = k(k-1)/2$ arises from a union bounding operation over the different sets of retained paths.

This reasoning can be extended to $M = 2^j$, in which case the path loss probability can be bounded above by the probability that the noise carries the transmitted signal point across a hyperplane through the center of the $(j+1)$-dimensional hypercube spanned by the $2M$ signal points after the extension step in the M-algorithm. This center has distance $d^2 = (j+1)4RE_b$ from the transmitted signal point, and therefore $\min(d_l^2) = (j+1)4RE_b$, and

$$\Pr(\text{CPL}) \leq \kappa Q\left(\sqrt{(\log_2 M + 1)2RE_b/N_0}\right). \tag{7.41}$$

As in the case of trellis codes, the factor κ is not only very difficult to calculate for $M > 2$, but also leads to an increasingly looser bound (7.41). We will therefore use the rough approximation

$$\Pr(\text{CPL}) \approx Q\left(\sqrt{(\log_2 M + 1)2RE_b/N_0}\right) \tag{7.42}$$

in our further arguments.

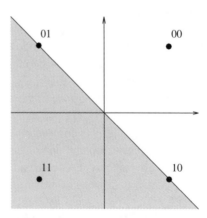

Figure 7.22 Signal constellation for the path loss probability for $M = 2$.

For ML decoding the block error probability is dominated by the codeword pair with the minimum Hamming distance d_H; that is,

$$\overline{P_{B,\mathrm{ML}}} \approx Q\left(\sqrt{d_H 2RE_b/N_0}\right). \tag{7.43}$$

Hence, for near optimality, we require that

$$\log_2 M + 1 \geq d_H. \tag{7.44}$$

If the correct path is retained at the beginning of the parity section of the code tree, the chance of selecting an erroneous final codeword is approximately that of making a decoding error in a block code with M codewords and minimum distance d_H. This is so since no more paths are discarded in the parity section of the code tree, and all discarded paths in the information section of the tree were erroneous paths.

$\Pr_{\mathrm{parity}}(\mathrm{CPL})$ is therefore also proportional to $Q\left(\sqrt{d_H 2RE_b/N_0}\right)$; that is, the path loss probability (7.42) in the information section of the tree is the important parameter. Now, the information section of the tree does not depend on the particular code used, and the only parameter of importance is the Hamming distance d_H, since it determines the size of M needed according to (7.44).

This analysis becomes impenetrable if M is not a power of 2. Even though we have carried out only an approximative analysis here, the precision of (7.44) is illustrated in the performance simulation figures presented earlier, for which the size of M required for near optimality was found to be $M = 4$ for the [7, 4] Hamming code, $M = 8$ for the [15, 7] BCH code, and $M = 64$ for the [23, 12] Golay code.

Clearly this procedure works well only if d_H is relatively small, that is, $d_H \lesssim 10$, due to the exponential growth of the complexity of the algorithm with d_H (from (7.44)). Block codes of relatively moderate size, however, have already sizable values of d_H, e.g., the [$n = 127, k = 64, d_h = 21$] BCH code already has $d_H = 21$, and would require $M = 2^{20}$. (A list of BCH codes and their parameters can be found, for example, in [37, Table 8-1-6].) In order to avoid this fast increase in complexity, we will use a sorting procedure to concentrate the reliable symbols at the beginning of the code tree. Let us start with the metric of codeword $\mathbf{x}^{(i)}$, given by (see Chapter 6)

$$J^{(i)} = |y - x^{(i)}|^2 = \sum_{r=1}^{n} |y_r - x_r^{(i)}|^2, \tag{7.45}$$

where $\mathbf{y}$ is the vector of received symbols from the channel. Equation (7.45) can be simplified to

$$J^{(i)} = -\sum_{r=1}^{n} y_r x_r^{(i)} \tag{7.46}$$

by neglecting terms common to all codewords. Recalling that we assume BPSK modulation, that is, $x_r^{(i)} \in \{-\sqrt{RE_b}, +\sqrt{RE_b}\}$, we note that the metric increments can be ordered according to their contribution to the total metric by ordering the received samples y_r according to their absolute values $|y_r|$. Let $\mathbf{y}' = \Pi(\mathbf{y})$ be the ordered set of received samples such that $|y_1'| \geq |y_2'| \geq \cdots \geq |y_n'|$, where $\Pi(\cdot)$ is the permutation of $\mathbf{y}$ that achieves this ordering. Letting $\mathbf{x}^{(i)'} = \Pi(\mathbf{x}^{(i)})$ we see that

$$J^{(i)} = J^{(i)'} = \sum_{r=1}^{n} y_r' x_r^{(i)'}, \tag{7.47}$$

where the unreliable samples, which carry a smaller contribution to $J^{(i)'}$, now come last.

This reordering presents us with one additional difficulty: The original code tree can no longer be used for decoding. However, we will show that a new code tree corresponding to the reordered code can be constructed in $O(kn)$ operations. We are using this altered code tree in the following amplitude-ordered M-algorithm:

Step 1.   Order the received symbols y_r according to their amplitudes. This generates a permutation $\mathbf{y}' = \Pi(\mathbf{y})$ of the received block of symbols $\mathbf{y}$. This ordering can be done in $O(n \log n)$ steps (for sorting algorithms and their complexity see [38]).

Step 2.   Generate the permuted encoder matrix $\mathbf{G}' = \Pi(\mathbf{G})$ by applying $\Pi(\cdot)$ to every row of $\mathbf{G}$. The codewords $\mathbf{x}'$ of $\mathbf{G}'$ correspond to the reordered received symbols $\mathbf{y}'$.

Step 3.   Generate the pseudosystematic encoder matrix $\mathbf{G}_p'$ (see below), which can be done in $O(kn)$ steps. $\mathbf{G}_p'$ is used to generate the code tree for the reordered code.

Step 4.   Apply the M-algorithm, deleting paths as needed to keep a maximum of M paths in the tree.

Step 5.   At the end of the tree, select the path $\hat{\mathbf{x}}'$ with the largest metric J' as the decoded codeword.

Step 6.   Apply the inverse permutation to obtain $\hat{\mathbf{x}} = \Pi^{-1}(\hat{\mathbf{x}}')$. The decoded information bits are now given by $\hat{\mathbf{u}} = \mathbf{G}^{-1}\hat{\mathbf{x}}$. If the original encoder $\mathbf{G}$ was in its systematic form, this step is trivially taking the first k entries of $\hat{\mathbf{x}}$.

The complexity of this ordering M-algorithm is $O(n \log n) + O(kn) + O(M)$ and is dominated by generating the reordered tree as long as $M < kn$. Figure 7.23 shows the BER of the moderate size $[63, 36, 11]$ BCH code using the reordering M-algorithm. The performance is nearly 2 dB better than hard-decision decoding, and hence close to optimal. The value of M used is 64. Unfortunately, an analysis of the size of M required for near-optimal performance is not available at this point, and it is not known whether near-optimal performance can be attained with a complexity that is not exponential in k for very large codes.

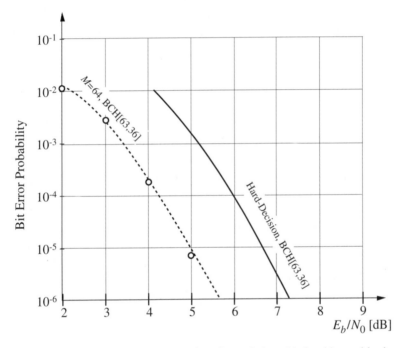

Figure 7.23 Simulation results using the ordering M-algorithm with the $[63, 36, 11]$ BCH using $M = 64$.

We have not yet explained the generation of the code tree for the ordered code, generated by $\mathbf{G}'$, and we will do so now. Most probably, $\mathbf{G}'$ will not be in systematic form, and we will generate a sequence of equivalence transformation to $\mathbf{G}'$ and generate a *pseudosystematic* encoder matrix $\mathbf{G}'_p$, which has k systematic output bits, defined as $x_r = u_r$, concentrated as far to the left as possible. The algorithm is very similar to the algorithm used to generate a systematic generator matrix for a linear block code [3] with the major difference that column permutations are not allowed. The systematic output bits can then be used as independent branching choices, and the remaining channel symbols are parity symbols and correspond to nondiverging sections in the tree.

Starting with an example, let

$$
\mathbf{G}' = \begin{bmatrix} 1 & 1 & 0 & 0 & 1 & 0 & 1 \\ 0 & 1 & 0 & 1 & 0 & 0 & 1 \\ 1 & 0 & 1 & 0 & 0 & 0 & 1 \\ 1 & 1 & 0 & 0 & 0 & 1 & 0 \end{bmatrix} \tag{7.48}
$$

be the generator matrix of the systematic [7, 4] Hamming code after the permutation $\Pi(\mathbf{G}) = (7, 6, 3, 2, 1, 4, 5)$. Applying only row operations, that is, permutations of rows and additions of one row to another,[4] we transform $\mathbf{G}'$ into

$$
\mathbf{G}'_p = \begin{bmatrix} 1 & 0 & 0 & 1 & 0 & 0 & 1 \\ 0 & 1 & 0 & 1 & 0 & 1 & 1 \\ 0 & 0 & 1 & 1 & 0 & 1 & 0 \\ 0 & 0 & 0 & 0 & 1 & 1 & 1 \end{bmatrix}, \tag{7.49}
$$

which has $k = 4$ systematic output bits in positions $1, 2, 3, 5$. The tree representation of $\mathbf{G}'_p$ is shown in Figure 7.24. It differs from that of the systematic code (Figure 7.18) in that the information portion and the parity portion of the tree are no longer separate.

As can be seen now, this procedure moves the decisions in the tree as far left as possible. But the leftmost portion of the ordered received sequence is where the large metric contributions are made, and our algorithm concentrates the decisions in this area.

The following is the general form of the algorithm to generate the pseudosystematic generator matrix $\mathbf{G}'_p$: Let the generator matrix be given by

$$
\mathbf{G}' = \begin{bmatrix} g'_1 \\ g'_2 \\ \vdots \\ g'_k \end{bmatrix}, \tag{7.50}
$$

where $g'_r = (g'_{r1}, g'_{r2}, \ldots, g'_{rn})$ is the the rth row of $\mathbf{G}'$.

Step 1.   Let $r = 1$ and $s = 0$.

Step 2.   Permute the rows r through n such that the $(r + s)$th entry of g'_r equals 1, i.e., such that $g'_{r(r+s)} = 1$. Now generate

---

[4] Add row 1 to rows 3 and 4, then add row 2 to rows 1 and 3, and, finally, add row 4 to rows 1 and 3.

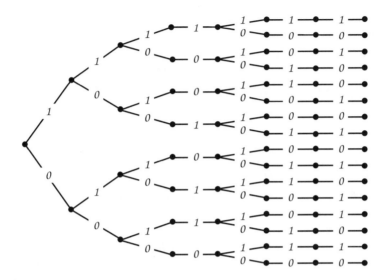

Figure 7.24 Tree representation of the reordered code with pseudosystematic generator matrix $\mathbf{G}'_p$.

$$
\mathbf{G}'' = \begin{bmatrix}
 & & 0 & \tilde{g}''_1 \\
 & & 0 & \vdots \\
 & \mathbf{G}'_p & 1 & \tilde{g}''_r \\
 & & 0 & \vdots \\
 & & 0 & \tilde{g}''_k
\end{bmatrix}, \tag{7.51}
$$

by adding (modulo 2) g'_r to g'_i if $g'_{i(r+s)} = 1$; that is, $g''_i = g'_r + g'_i$ and $\tilde{g}''_r$ is the partial row r. This produces the column

$$
(\underbrace{0 \cdots 0}_{r-1 \text{ zeros}} 10 \cdots 0)^T
$$

in position $r + s$. $\mathbf{G}'_p$ is the part already in pseudosystematic form.

Step 3.   If no such permutation as required in Step 2 exists, set $s = s + 1$.

Step 4.   If $r < k$, $\mathbf{G}'' \to \mathbf{G}'$, and go to Step 2; else output $\mathbf{G}' = \mathbf{G}'_p$, which is the pseudosystematic encoder matrix.

As a final thought we would like to dwell on the apparent difference between the direction of many of our theoretical results and the practical decoding examples in this section. It is noteworthy, and maybe somewhat disconcerting, that the M-algorithm requires no knowledge of the algebraic code structure nor any knowledge of the minimal trellis of the code. And

yet, it seems, that it is a very cost-effective algorithm in terms of decoding complexity versus performance (at least for codes of moderate length). It seems that the intricate structure of a code plays only a minor role in its application, since some of the the algorithms with which these codes are decoded successfully require none of that information.

References

[1] H. Imai et al., *Essentials of Error-Control Coding Techniques*, Academic Press, New York, 1990.

[2] R. McEliece, *The Theory of Information and Coding, Encyclopedia of Mathematics and Its Applications*, Addison-Wesley, Reading, MA, 1977.

[3] F. J. MacWilliams and N. J. A. Sloane, *The Theory of Error Correcting Codes*, North-Holland, Amsterdam, 1988.

[4] E. R. Berlekamp, *Algebraic Coding Theory*, McGraw-Hill, New York, 1968.

[5] J. L. Massey, "Shift-register synthesis and BCH decoding," *IEEE Trans. Inform. Theory*, Vol. IT-15, pp. 122–127, 1969.

[6] G. D. Forney, Jr., "Generalized minimum distance decoding," *IEEE Trans. Inform. Theory*, Vol. IT-12, pp. 125–131, 1966.

[7] E. J. Weldon, Jr., "Decoding binary block codes on q-ary output channels," *IEEE Trans. Inform. Theory*, Vol. IT-17, pp. 713–718, 1971.

[8] D. Chase, "A class of algorithms for decoding block codes with channel measurement information," *IEEE Trans. Inform. Theory*, Vol. IT-18, pp. 170–182, 1972.

[9] K. R. Matis and J. W. Modestino, "Reduced-state soft-decision trellis decoding of linear block codes," *IEEE Trans. Inform. Theory*, Vol. IT-28, pp. 61–68, 1982.

[10] J. Snyders and Y. Be'ery, "Maximum likelihood soft decoding of binary block codes and decoders for the Golay codes," *IEEE Trans. Inform. Theory*, Vol. IT-35, pp. 963–975, 1989.

[11] F. Hemmati, "Closest coset decoding of $|u|u + v|$ codes," *IEEE J. Select. Areas Commun.*, Vol. SAC-7, pp. 982–988, 1989.

[12] N. J. C. Lous, P. A. H. Bours, and H. C. A. van Tilborg, "On maximum likelihood soft-decision decoding of binary linear codes," *IEEE Trans. Inform. Theory*, Vol. IT-39, pp. 197–203, 1993.

[13] Y. S. Han, C. R. P. Hartman, and C. C. Chen, "Efficient priority-first search maximum-likelihood soft-decision decoding of linear block codes," *IEEE Trans. Inform. Theory*, Vol. IT-39, pp. 1514–1523, 1993.

[14] A. Vardy and Y. Be'ery, "More efficient soft decoding of the Golay codes," *IEEE Trans. Inform. Theory*, Vol. IT-37, pp. 667–672, 1991.

[15] A. Vardy and Y. Be'ery, "Maximum likelihood soft decoding of BCH codes," *IEEE Trans. Inform. Theory*, Vol. IT-40, pp. 546–554, 1994.

[16] D. J. Taipale and M. J. Seo, "An efficient soft-decision Reed-Solomon decoding alorithm," *IEEE Trans. Inform. Theory*, Vol. IT-40, pp. 1130–1139, 1994.

[17] T. Taneko, T. Nishijima, H. Inazumi, and S. Hirasawa, "Efficient maximum likelihood decoding of linear block codes with algebraic decoder," *IEEE Trans. Inform. Theory*, Vol. IT-40, pp. 320–327, 1994.

[18] M. P. C. Fossorier and S. Lin, "Soft-decision decoding of linear block codes based on ordered statistics," *IEEE Trans. Inform. Theory*, Vol. IT-41, pp. 1379–1396, 1995.

[19] S. Lin and D. J. Costello, Jr., *Error Control Coding*, Prentice-Hall, Englewood Cliffs, NJ, 1983.

[20] R. E. Blahut, *Principles and Practice of Information Theory,* Addison-Wesley, Reading, MA, 1987.

[21] G. C. Clark and J. B. Cain, *Error-Correction Coding for Digital Communications,* Plenum Press, New York, 1983.

[22] L. R. Bahl, J. Cocke, F. Jelinek, and J. Raviv, "Optimal decoding of linear codes for minimizing symbol error rate," *IEEE Trans. Inform. Theory*, Vol. IT-20, pp. 284–287, 1974.

[23] J. K. Wolf, "Efficient Maximum-likelihood decoding of linear block codes using a trellis," *IEEE Trans. Inform. Theory*, Vol. IT-24, No. 1, pp. 76–80, 1978.

[24] J. L. Massey, "Foundations and methods of channel coding," *Proc. Int. Conf. Inform. Theory Systems, NTG-Fachber.* Vol. 65, pp. 148–157, 1978.

[25] R. J. McEliece, "On the BCJR trellis for linear block codes," unpublished manuscript, 1994.

[26] G. Strang, *Linear Algebra and Its Applications*, Harcourt Brace Jovanovich, San Diego, 1988.

[27] S. Dolinar et al., "Trellis complexity of linear block codes," unpublished manuscript, 1995.

[28] T. Kasami et al., "On the optimum bit orders with respect to the state complexity of trellis diagrams for binary linear codes," *IEEE Trans. Inform. Theory*, Vol. IT-39, pp. 242–245, 1993.

[29] T. Kasami et al., "On the trellis structure of block codes," *IEEE Trans. Inform. Theory*, Vol. IT-39, pp. 1057–1064, 1993.

[30] G. D. Forney, "Coset codes. II: Binary lattices and related codes," *IEEE Trans. Inform. Theory*, Vol. IT-34, No. 5, pp. 1152–1187, 1988.

[31] J. H. Conway and N. J. A. Sloane, *Sphere Packings, Lattices and Groups*, Springer-Verlag, New York, 1988.

[32] G. R. Lang and F. M. Longstaff, "A Leech lattice modem," *IEEE J. Select. Areas Commun.*, Vol. SAC-7, No. 6, pp. 968–973, 1989.

[33] J. H Conway and N. J. A. Sloane, "Decoding techniques for codes and lattices, including the Golay code and the Leech lattice," *IEEE Trans. Inform. Theory*, Vol. IT-32, pp. 41–50, 1986.

[34] B. Jelicic, C. Schlegel, and S. Roy, "Soft decision decoding of block codes using the M-algorithm," unpublished manuscript, 1996.

[35] L. Ma and C. Schlegel, "Efficient algorithmic decoding of block codes," unpublished manuscript, 1996.

[36] L. Wei and C. Schlegel, "Synchronous DS-SSMA with improved decorrelating decision-feedback multiuser detection," *IEEE Trans. Veh. Technol.*, Vol. VT-43, No. 3, pp. 767–772, 1994.

[37] J. G. Proakis, *Digital Communications*, 3rd ed., McGraw-Hill, New York, 1995.

[38] D. E. Knuth, *The Art of Computer Programming. Vol. 3: Sorting and Searching*, Addison-Wesley, Reading, MA, 1973.

CHAPTER 8

TURBO CODES

Lance C. Perez

8.1 INTRODUCTION

In previous chapters, several encoding and decoding techniques that have achieved significant gains in terms of power efficiency and bandwidth efficiency were discussed. Though this has resulted in widespread application of forward error correction (FEC) coding techniques, the quest for practical techniques that achieve near-capacity performance has met with limited success, particularly in power-limited applications. This is, of course, not totally unexpected since the basic theorems of information theory imply that capacity can only be achieved through the use of coding schemes with infinite complexity.

Historically, coding gains in practical systems have been limited by the technology available with which to implement them. That is, at the time of the earliest applications of FEC coding better codes were available, but it was not feasible to implement them. As technology has advanced, so has the ability to implement complex coding schemes as demonstrated by the so-called big Viterbi decoder (BVD) built by the Jet Propulsion Laboratory that decodes rate-$1/n$ codes with 2^{14} states. As discussed in Chapter 1, the use of more complex coding schemes did reduce the gap between real system performance and theoretical limits. However, the gains of these complex codes were not as dramatic as expected, and it appeared that a law of diminishing returns was manifesting itself.

In 1993 a new coding and decoding scheme, dubbed Turbo codes by its discoverers, was reported that achieves near-capacity performance on the additive white Gaussian noise channel. The discovery of Turbo codes and the near-capacity performance has stimulated a flurry of research effort to fully understand this new coding scheme [1–27]. From this emerged two fundamental questions regarding Turbo codes. First, does the iterative decoding scheme presented always converge to the optimal solution [18, 24, 25]? Second, assuming optimal or near-optimal decoding, why do the Turbo codes perform so well?

The simulated performance of a Turbo code is shown in Figure 8.1 along with simulation results for a rate $R = 1/2$, memory $\nu = 14$, convolutional code. The comparison of these simulation results raises two issues regarding the performance of Turbo codes. First, what is it that allows Turbo codes to achieve a bit error rate (BER) of 10^{-5} at a signal-to-noise ratio (SNR) of only 0.7 dB? Second, what causes the "error floor?" that is, the flattening of the performance curve, for moderate to high SNRs?

In this chapter, the fundamentals of this new coding and decoding scheme are discussed with the emphasis on explaining the spectacular per-

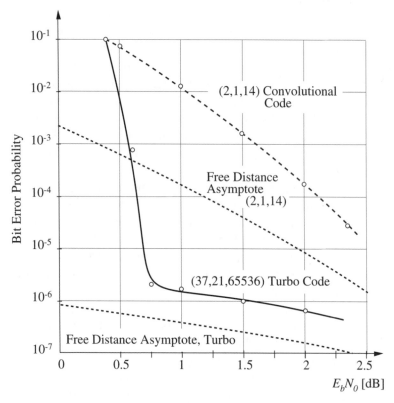

Figure 8.1 Simulated performance of the original Turbo code and the code's free-distance asymptote and the simulated performance of the (2, 1, 14) MFD convolutional code.

formance of Turbo codes. Here, it is endeavored to explain the performance of Turbo codes, and thus address the two issues mentioned above,
in terms of the code's distance spectrum. The iterative algorithm used to
decode Turbo codes is straightforward to implement, but most of its more
detailed properties, such as convergence, are unknown at this time. For
this reason, the discussion of the decoder in this chapter is limited to a brief
description of the iterative algorithm. The interested reader is referred to
the literature [18, 24, 25] for current results regarding the decoder.

To explain their performance in terms of the distance spectrum, the
codeword structure of Turbo codes will be examined in detail. The free
distance is defined to be the minimum Hamming weight of all possible
codewords, and the error coefficient is the number, or multiplicity, of free-
distance codewords. The goal is to elucidate the key structural properties
that result in the near-capacity performance of Turbo codes at BERs around
10^{-5}. As will be seen, this effort leads to an interpretation that applies to
Turbo codes and lends insight into designing codes in general. Throughout,
Turbo codes are compared to a maximum free-distance rate $R = 1/2$, memory $v = 14$, that is, $(2, 1, 14)$, convolutional code to emphasize the differences in performance and structure. Techniques for analyzing the performance of Turbo codes using transfer functions may be found in [5, 8, 16].

8.2   CODEWORD STRUCTURE OF TURBO CODES

In order to find the free distance of a Turbo code, it is necessary to understand the basic structure of the encoder and the resulting codewords.
A Turbo encoder consists of the *parallel concatenation* of two or more,
usually identical, rate-1/2 encoders, realized in systematic feedback form,
and a pseudorandom interleaver. This encoder structure is called a parallel
concatenation because the two encoders operate on the same *set* of input
bits, rather than one encoding the output of the other. A block diagram of
a Turbo encoder with two constituent convolutional encoders is shown in
Figure 8.2. For the remainder of the chapter, only Turbo encoders with two
constituent convolutional encoders are considered, though the conclusions
are easily extended to the case where three or more encoders are used [15].

The interleaver is used to permute the input bits such that the two
encoders are operating on the same *set* of input bits, but different input
sequences. Thus, the first encoder receives the input bit u_r and produces
the output pair $(u_r, v_r^{(1)})$, and the second encoder receives the input bit u_r'
and produces the output pair $(u_r', v_r^{(2)})$. The input bits are grouped into

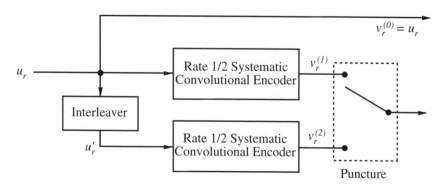

Figure 8.2 Block diagram of a Turbo encoder with two constituent encoders and an optional puncturer.

finite-length sequences whose length, N, equals the size of the interleaver. Since both the encoders are systematic and operate on the same set of input bits, it is only necessary to transmit the input bits once and the overall code has rate $1/3$. In order to increase the overall rate of the code to $1/2$, the two parity sequences $\mathbf{v}^{(1)}$ and $\mathbf{v}^{(2)}$ can be punctured by alternately deleting $v^{(1)}$ and $v^{(2)}$. We will refer to a Turbo code whose constituent encoders have parity-check polynomials h_0 and h_1, expressed in either octal or D transform notation, and whose interleaver is of length N as an (h_0, h_1, N) Turbo code.

For example, consider the Turbo encoder in Figure 8.2, where each constituent encoder is a $(2, 1, 2)$ encoder with parity-check polynomials $h_0(D) = 1 + D^2$ and $h_1(D) = D$. Assume a pseudorandom interleaver of size $N = 16$ bits that generates a $(1 + D^2, D, 16)$ Turbo code. The interleaver is realized as a 4×4 matrix filled sequentially, row by row, with the input bits u_r. Once the interleaver has been filled, the input bits to the second encoder, u'_r, are obtained by reading the interleaver in a pseudorandom manner until each bit has been read once and only once. The pseudorandom nature of the interleaver in this example is represented by a permutation $\Pi_{16} = \{15, 10, 1, 12, 2, 0, 13, 9, 5, 3, 8, 11, 7, 4, 14, 6\}$, which implies $u'_0 = u_{15}$, $u'_1 = u_{10}$, and so on.

If $\mathbf{u} = \{u_0, \ldots, u_{15}\} = \{1, 0, 0, 0, 1, 0, 0, 0, 0, 0, 0, 0, 1, 0, 1, 0\}$ is the input sequence and the interleaver is represented by the permutation Π_{16}, then the sequence $\mathbf{u}' = \Pi_{16}(\mathbf{u}) = \{0, 0, 0, 1, 0, 1, 0, 0, 0, 0, 0, 0, 0, 1, 1, 0\}$ is the input to the second encoder. The trellis diagrams for both constituent encoders with these inputs are shown in Figure 8.3. The corresponding unpunctured parity sequences are $\mathbf{v}^{(1)} = \{0, 1, 0, 1, 0, 0, 0, 0, 0, 0, 0, 0, 0, 1,$

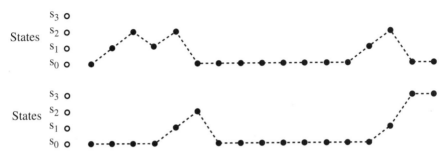

Figure 8.3 Examples of detours in the constituent encoders.

$0, 0\}$ and $\mathbf{v}^{(2)} = \{0, 0, 0, 0, 1, 0, 0, 0, 0, 0, 0, 0, 0, 0, 1, 1\}$. The resulting codeword has Hamming weight $d = w(\mathbf{u}) + w(\mathbf{v}^{(1)}) + w(\mathbf{v}^{(2)}) = 4+3+3 = 10$ without puncturing, where $w(\mathbf{u})$ is the Hamming weight of the sequence $\mathbf{u}$. If the code is punctured beginning with $v_0^{(1)}$, then the resulting codeword has weight $d = 4+3+2 = 9$. If, on the other hand, the puncturing begins with $v_0^{(2)}$, then the punctured codeword has Hamming weight $4+0+1 = 5$.

Finding the free distance of a Turbo code is complicated by the fact that Turbo encoders are time varying due to the interleaver. That is, if $\tilde{\mathbf{u}} = D\mathbf{u}$, where D is the delay operator, then $\tilde{\mathbf{v}}^{(1)} = D\mathbf{v}^{(1)}$, but $\tilde{\mathbf{u}}' \neq D\mathbf{u}'$ and thus $\tilde{\mathbf{v}}^{(2)} \neq D\mathbf{v}^{(2)}$ with high probability. Note that we only consider delays of a finite-length sequence $\mathbf{u}$ for which no ones are lost and not cyclic shifts. Continuing the example, if $\tilde{\mathbf{u}} = D\mathbf{u}$, then $\tilde{\mathbf{u}}' = \Pi_{16}(\tilde{\mathbf{u}}) = \{1, 0, 1, 0, 0, 0, 1, 0, 1, 0, 0, 0, 0, 0, 0, 0\}$ and the second parity sequence is $\tilde{\mathbf{y}}^{(2)} = \{0, 1, 0, 0, 0, 0, 0, 1, 0, 0, 0, 0, 0, 0, 0, 0\}$. Thus, time-shifting the input bits results in codewords that differ in both bit position and overall Hamming weight! The variation in the weights of codewords corresponding to time-shifted input sequences is magnified by puncturing.

This simple example illustrates several salient points concerning the structure of the codewords in a Turbo code. First, because the pseudorandom interleaver permutes the input bits, the two input sequences $\mathbf{u}$ and $\mathbf{u}'$ are almost always different, though of the same weight, and the two encoders will (with high probability) produce parity sequences of different weights. Second, it is easily seen that a codeword may consist of a number of distinct detours in each encoder. Note that since the constituent encoders are realized in systematic feedback form, a nonzero sequence is required to return to the all-zero state and thus all detours are associated with information sequences of weight 2 or greater. Finally, with a pseudorandom interleaver it is highly unlikely that both encoders will be returned to the all-zero state at the end of the codeword even when the last v bits of the input sequence $\mathbf{u}$ are chosen to force the first encoder back to the all-zero state.

If neither encoder is forced to the all-zero state (i.e., no tail is used), then the sequence consisting of $N - 1$ zeros followed by a 1 is a valid input sequence $\mathbf{u}$ to the first encoder. For some interleavers, this $\mathbf{u}$ will be permuted to itself and $\mathbf{u}'$ will be the same sequence. In this case, both $\mathbf{v}^1$ and $\mathbf{v}^2$ have weight zero and the overall weight of the codeword and the free distance of the code is 2, regardless of whether or not puncturing is used. Note that this codeword is caused by an information sequence of weight 1. For this particular encoder, any information sequence of weight 1 results in a parity sequence of alternating 1's and 0's. Consequently, even if $\mathbf{u}'$ is not the same as $\mathbf{u}$, puncturing can result in a weight 2 codeword. For systematic feedback encoders, forcing the first encoder to return to the all-zero state insures that every information sequence will have at least weight two. For this reason, it is common to assume that the first encoder is forced to return to the all-zero state.

The ambiguity of the final state of the second encoder has been shown by simulation to result in negligible performance degradation [27, 9] for large interleavers. For these reasons, it will be assumed for the remainder of this chapter that the first encoder is forced to return to the all-zero state and that the final state of the second encoder is unknown. Special interleaver structures that result in both encoders returning to the all-zero state are discussed in [11, 12, 20].

8.3   ITERATIVE DECODING OF TURBO CODES

It is clear from the discussion of the codeword structure of Turbo codes that the state-space of these codes is too large to perform optimal decoding. To overcome this, the discoverers of Turbo codes proposed a novel iterative decoder based on the maximum a posteriori (MAP) decoding algorithm described in Section 6.6. Empirical evidence suggests that this decoding algorithm performs remarkably well and may converge to the optimal decoding solution. The convergence properties of the iterative algorithm remain an interesting open research question and are discussed in the literature. For the current discussion, a description of the algorithm is sufficient.

In Section 6.6 it was shown that the maximum a posteriori decoder computes the a posteriori probability $\Pr(u_r = u | \mathbf{y})$ conditioned on the received sequence $\mathbf{y}$ (see equation (6.44)). The iterative Turbo decoder makes use of these a posteriori probabilities in the form of a log-likelihood ratio given by

$$L(u_r) = \log \frac{\Pr(u_r = 1|\mathbf{y})}{\Pr(u_r = 0|\mathbf{y})},$$
(8.1)

which, from (6.38) and (6.44), is given by

$$L(u_r) = \log \frac{\displaystyle\sum_{(i,j)\in A(u_r=1)} \gamma_r(j, i)d_{r-1}(i)\beta_r(j)}{\displaystyle\sum_{(i,j)\in A(u_r=0)} \gamma_r(j, i)d_{r-1}(i)\beta_r(j)}.$$
(8.2)

The joint conditional probability $\gamma_r(j, i)$ may be expressed as

$$\gamma_r(j, i) = p_{ij}\Pr(y_r^{(0)}, y_r^{(m)}|u_r, v_r^{(m)})$$
(8.3)

through the use of (6.43), where $y_r^{(0)}$ is the received systematic bit and $y_r^{(m)}$, $m = 1, 2$, is the received parity bit corresponding to the mth constituent encoder. For systematic codes, the last term of (8.3) may be factored as

$$\Pr(y_r^{(0)}, y_r^{(m)}|u_r, v_r^{(m)}) = \Pr(y_r^{(0)}|u_r)\Pr(y_r^{(m)}|v_r^{(m)}),$$
(8.4)

since the received systematic sequence and the received parity sequence are independent of each other. Finally, substituting (8.3) and (8.4) into (8.2) and factoring yields

$$L(u_r) = \log \frac{\displaystyle\sum_{(i,j)\in A(u_r=1)} \Pr(y_r^{(m)}|v_r^{(m)})\alpha_{r-1}(i)\beta_r(j)}{\displaystyle\sum_{(i,j)\in A(u_r=0)} \Pr(y_r^{(m)}|v_r^{(m)})\alpha_{r-1}(i)\beta_r(j)}$$

$$+ \log \frac{\Pr(u_r = 1)}{\Pr(u_r = 0)} + \log \frac{\Pr(y_r^{(0)}|u_r = 1)}{\Pr(y_r^{(0)}|u_r = 0)}$$
(8.5)

$$= \Lambda_{e,r}^{(m)} + \Lambda_r + \Lambda_s,$$
(8.6)

where $\Lambda_{e,r}^{(m)}$ is called the *extrinsic information* from the mth decoder, Λ_r is the *a priori* log-likelihood ratio of the systematic bit u_r, and Λ_s is the log-likelihood ratio of the *a posteriori* probabilities of the systematic bit. Having developed 8.6, we are now ready to formulate the Turbo decoding algorithm.

A block diagram of an iterative Turbo decoder is shown in Figure 8.4, where each MAP decoder corresponds to a constituent encoder. The interleavers are identical to the interleavers in the Turbo encoder and are used to reorder the sequences so that each decoder is properly synchronized.

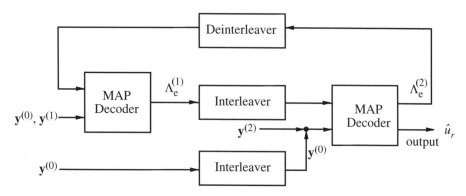

Figure 8.4 Block diagram of a Turbo decoder with two constituent decoders.

For the first iteration, the first decoder computes the log-likelihood ratio of equation (8.6) with $\Lambda_r = 0$, since u_r is equally likely to be a 0 or a 1, using the received sequences $\mathbf{y}^{(0)}$ and $\mathbf{y}^{(1)}$. The second decoder now computes the log-likelihood ratio of equation (8.6) based on the received sequences $\mathbf{y}^{(0)}$ and $\mathbf{y}^{(2)}$ (suitably reordered).

Notice, however, that the second decoder has available an estimate of the *a posteriori* probability of u_r from the first decoder, namely $L^{(1)}(u_r)$. The second decoder may consider this as the *a priori* probability of u_r in (8.6) and compute

$$
\begin{aligned}
L^{(2)}(u_r) &= \Lambda_{e,r}^{(2)} + L^{(1)}(u_r) + \Lambda_s \\
&= \Lambda_{e,r}^{(2)} + \Lambda_{e,r}^{(1)} + \Lambda_r^{(1)} + \Lambda_s + \Lambda_s
\end{aligned}
\tag{8.7}
$$

as the new log-likelihood ratio. Close examination of (8.7) reveals that by passing $L^{(1)}(u_r)$, the second decoder is given $\Lambda_r^{(1)}$, the previous estimate of the *a priori* probability, which is clearly unnecessary. In addition, it is easily seen that as the Turbo decoder continues to iterate, the log-likelihood ratio accumulates Λ_s and the systematic bit becomes overemphasized. To prevent this, the second decoder subtracts $\Lambda_r^{(1)}$ and Λ_s from the information passed from the first decoder resulting in

$$
L^{(2)}(u_r) = \Lambda_{e,r}^{(2)} + \Lambda_{e,r}^{(1)} + \Lambda_s.
\tag{8.8}
$$

From (8.8), we see that critical information that is passed between the decoders is in fact the extrinsic information. On subsequent iterations, the first decoder uses $\Lambda_{e,r}^{(2)}$ to compute the *a posteriori* probability of u_r. This process continues until a desired performance is achieved at which point a final decision is made by comparing the final log-likelihood ratio to the threshold 0.

The extrinsic information is a reliability measure of each component decoder's estimate of the transmitted sequence based on the corresponding received component parity sequence and is essentially independent of the received systematic sequence. Since each component decoder uses the received systematic sequence directly, the extrinsic information allows the decoders to share information without significant error propagation. The efficacy of this technique can be seen in Figure 8.5, which shows the performance of the original (37,21,65536) Turbo code as a function of the decoder iterations. It is impressive that the performance of the code with iterative decoding continues to improve up to 18 iterations.

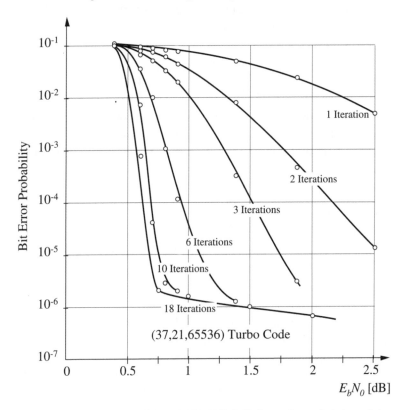

Figure 8.5 Performance of the (37,21,65536) Turbo code as function of the number of decoder iterations.

8.4   PERFORMANCE BOUNDS

In Chapter 5 the bounds on the performance of convolutional codes were developed, assuming an infinite trellis. The result is the bound of equation

(5.10), which gives the average information bit error rate *per unit time*. To clarify the distinction between Turbo codes and convolutional codes, it is useful to consider both codes as block codes and to redevelop the performance bounds from this point of view. To this end, the input sequences are restricted to length N, where N corresponds to the size of the interleaver in the Turbo encoder. With finite-length input sequences of length N, a $(2, 1, \nu)$ convolutional code may be viewed as a block code with 2^N codewords of length $2(\nu + N)$. The last 2ν bits are due to the ν bit tail used to force the encoder back to the all-zero state.

The bit error rate (BER) performance of a finite-length convolutional code with maximum-likelihood (ML) decoding on an additive white Gaussian noise (AWGN) channel with an SNR of E_b/N_0 is bounded above by

$$P_b \le \sum_{i=1}^{2^N} \frac{w_i}{N} Q\left(\sqrt{d_i \frac{RE_b}{2N_0}}\right),$$

where w_i and d_i are the information weight and total Hamming weight, respectively, of the ith codeword. Collecting codewords of the same total Hamming weight and defining the average information weight per codeword as

$$\tilde{w}_d = \frac{W_d}{N_d},$$

where W_d is the total information weight of all codewords of weight d and N_d is the number, or multiplicity, of codewords of weight d, yields

$$P_b \le \sum_{d=d_{\text{free}}}^{2(\nu+N)} \frac{N_d \tilde{w}_d}{N} Q\left(\sqrt{d \frac{RE_b}{2N_0}}\right),$$

where d_{free} is the free distance of the code. In this development, N_d includes codewords due to multiple detours for $d \ge 2d_{\text{free}}$.

If a convolutional code has N_d^0 codewords of weight d caused by information sequences $\mathbf{u}$ whose first one occurs at time 0, then it also has N_d^0 codewords of weight d caused by the information sequences $D\mathbf{u}$, N_d^0 codewords of weight d caused by the information sequences $D^2\mathbf{u}$, and so on. Thus, as the length of the information sequences increases, we have

$$\lim_{N \to \infty} \frac{N_d}{N} = N_d^0$$

and

$$\lim_{N \to \infty} \tilde{w}_d = \lim_{N \to \infty} \frac{W_d}{N_d} = \frac{W_d^0}{N_d^0} \triangleq \tilde{w}_d^0,$$

where W_d^0 is the total information weight of all codewords with weight d that are caused by information sequences whose first one occurs at time 0. Thus, the bound on the BER of a convolutional code with ML decoding becomes

$$P_b \leq \sum_{d=d_{\text{free}}}^{2(v+N)} N_d^0 \tilde{w}_d^0 Q\left(\sqrt{d\frac{RE_b}{2N_0}}\right) = \sum_{d=d_{\text{free}}}^{2(v+N)} W_d^0 Q\left(\sqrt{d\frac{RE_b}{2N_0}}\right), \qquad (8.9)$$

which is the standard union bound for ML decoding as developed in chapter 5. For this reason, efforts to find good convolutional codes for use with ML decoders have focused on finding codes that maximize the free distance d_{free} and minimize the number of free-distance paths N_{free}^0 for a given rate and constraint length.

The performance of a Turbo code with ML decoding is also bounded by

$$P_b \leq \sum_{i=1}^{2^N} \frac{w_i}{N} Q\left(\sqrt{d_i\frac{RE_b}{2N_0}}\right),$$

where w_i and d_i are the information weight and total Hamming weight, respectively, of the ith codeword. However, in the Turbo encoder the pseudorandom interleaver maps the input sequence $\mathbf{u}$ to $\mathbf{u}'$ and the input sequence $D\mathbf{u}$ to a sequence $\mathbf{u}''$ that is different from $D\mathbf{u}'$ with very high probability. Thus, unlike convolutional codes, the input sequences $\mathbf{u}$ and $D\mathbf{u}$ produce different codewords with different Hamming weights.

Collecting codewords of the same total Hamming weight, the bound on the BER for Turbo codes becomes

$$P_b \leq \sum_{d=d_{\text{free}}}^{2(v+N)} \frac{N_d \tilde{w}_d}{N} Q\left(\sqrt{d\frac{RE_b}{2N_0}}\right). \qquad (8.10)$$

For Turbo codes with pseudorandom interleavers, $N_d \tilde{w}_d$ is much less than N for low-weight codewords. This is due to the pseudorandom interleaver that maps low-weight parity sequences in the first constituent encoder to high-weight parity sequences in the second constituent encoder. Thus, for low-weight codewords

$$\frac{\tilde{w}_d N_d}{N} \ll 1,$$

where

$$\frac{N_d}{N} \qquad (8.11)$$

is called the *effective multiplicity* of codewords of weight d.

For moderate and high signal-to-noise ratios, it is well known that the free-distance term in the union bound on the BER performance dominates the bound. Thus, for Turbo codes the asymptotic performance approaches

$$P_b \approx \frac{N_{\text{free}} \tilde{w}_{\text{free}}}{N} Q \left(\sqrt{d_{\text{free}} \frac{RE_b}{2N_0}} \right), \tag{8.12}$$

where N_{free} is the error coefficient and $\tilde{w}_{\text{free}}$ is the average weight of the information sequences causing free-distance codewords. The expression on the right side of equation (8.12) and its associated graph is called the *free-distance asymptote*, P_{free}, of a Turbo code.

An algorithm for finding the free distance of Turbo codes is described in [26]. This algorithm was applied to a Turbo code with the same constituent encoders, puncturing pattern, and interleaver size N as in [1] and a *particular* pseudorandom interleaving pattern. The parity-check polynomials for this code are $h_0 = D^4 + D^3 + D^2 + D + 1$ and $h_1 = D^4 + 1$, or $h_0 = 37$ and $h_1 = 21$ in octal notation. This $(37, 21, 65536)$ code was found to have $N_{\text{free}} = 3$ paths with weight $d_{\text{free}} = 6$. Each of these paths was caused by an input sequence of weight 2 and thus $\tilde{w}_{\text{free}} = 2$. Though this result was for a particular pseudorandom interleaver, it is true for most pseudorandom interleavers with $N = 65536$. This is consistent with the conclusions in [5] in which the performance of Turbo codes is averaged over all possible pseudorandom interleavers.

For this Turbo code, the free-distance asymptote is

$$P_{\text{free}} = \frac{3 \cdot 2}{65536} Q \left(\sqrt{6 \frac{0.5E_b}{2N_0}} \right),$$

where the rate loss due to the addition of a 4-bit tail to terminate the first encoder is ignored and

$$\frac{N_{\text{free}}}{N} = \frac{3}{65536}$$

is the effective multiplicity. The free-distance asymptote is plotted in Figure 8.1 along with simulation results for this code using the iterative decoding algorithm with 18 iterations.

From Figure 8.1 it can clearly be seen that the simulation results do approach the free-distance asymptote for moderate and high SNRs. Since the slope of the asymptote is essentially determined by the free distance of the code, it can be concluded that the error floor observed with Turbo codes is due to the fact that they have a relatively small free distance and consequently a relatively flat free-distance asymptote.

Further examination of equation (8.12) reveals that the manifestation of the error floor can be manipulated in two ways. First, increasing the length of the interleaver while preserving the free distance and the error coefficient will lower the asymptote without changing its slope by reducing the effective multiplicity. In this case, the performance curve of a Turbo code does not flatten out until higher SNRs and lower BERs are reached. Conversely, decreasing the interleaver size while maintaining the free distance and error coefficient results in the error floor being raised and the performance curve flattens at lower SNRs and higher BERs. This can be seen in the simulation results for the Turbo code with varying N shown in Figure 8.6. If the first constituent encoder is not forced to return to the all-zero state and the weight-2 codewords described at the end of Section 8.2 are allowed, then the error floor is raised to the extent that the code performs poorly even for large interleavers. Thus, one cannot completely disregard free distance when constructing Turbo codes.

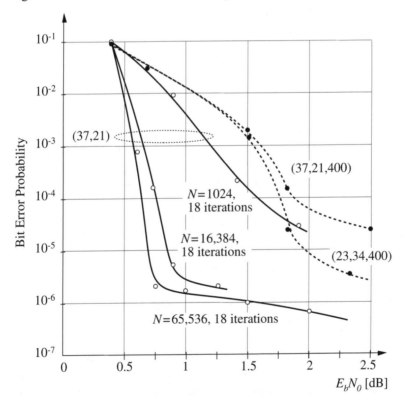

Figure 8.6 Simulations illustrating the effect of the interleaver size on the performance of a Turbo code.

If the size of the interleaver is fixed, then the "error floor" can be modified by increasing the free distance of the code while preserving the error coefficient. This has the effect of changing the slope of the free-distance asymptote. That is, increasing the free distance increases the slope of the asymptote, and decreasing the free distance decreases the slope of the asymptote. It has been shown in [7] and [26] that for a fixed interleaver size, choosing the feedback polynomial to be a primitive polynomial results in an increased free distance and thus a steeper asymptote. An argument to support the use of primitive polynomials in Turbo codes is presented in Section 8.8.

The role that the free distance and effective multiplicity play in determining the asymptotic performance of a Turbo code is further clarified by examining the asymptotic performance of a convolutional code. The free-distance asymptote of a convolutional code is given by the first term in the union bound of equation (8.9). The maximum free-distance $(2, 1, 14)$ code whose performance is shown in Figure 8.1 has $d_{\text{free}} = 18$, $N^0_{\text{free}} = 18$, and $W^0_{\text{free}} = 137$ [29]. Thus, the free-distance asymptote for this code is

$$P_{\text{free}} = 137 \; Q \left(\sqrt{18 \frac{0.5 E_b}{2 N_0}} \right),$$

which is also shown in Figure 8.1.

As expected, the free-distance asymptote of the $(2, 1, 14)$ code is much steeper than the free-distance asymptote of the Turbo code due to the increased free distance. However, because the effective multiplicity of the free-distance codewords of the Turbo code, given by equation (8.11), is much smaller than the multiplicity of the $(2, 1, 14)$ code, the two asymptotes do not cross until an SNR of $E_b/N_0 = 2.5$ dB. At this SNR, the BER of both codes is 10^{-6}, which is lower than the targeted BER of many practical systems. Thus, even though the $(2, 1, 14)$ convolutional code is asymptotically better than the $(37, 21, 65536)$ Turbo code, the Turbo code is better for the error rates at which many systems operate.

8.5   A TURBO CODE WITH A RECTANGULAR INTERLEAVER

To emphasize the importance of using a pseudorandom interleaver with Turbo codes, we now consider a Turbo code with a rectangular interleaver. The same constituent encoders and puncturing pattern as in [1] are used in conjunction with a 120×120 rectangular interleaver. This rectangular interleaver is realized as a 120×120 matrix into which the information sequence **u** is written row by row. The input sequence to the second encoder **u′** is then obtained by reading the matrix column by column. A 120×120 rectangular interleaver implies an interleaver size of $N = 14{,}400$ and thus this is a $(37, 21, 14400)$ Turbo code.

Using the algorithm described in [26], this code was found to have a free distance of $d_{free} = 12$ with a multiplicity of $N_{free} = 28{,}900$! For this code, each of the free-distance paths is caused by an information sequence of weight 4, so $\tilde{w}_{free} = 4$. The free-distance asymptote for this code is thus

$$P_{free} = \frac{28{,}900 \cdot 4}{14{,}400} Q\left(\sqrt{12 \frac{0.5E_b}{N_0}}\right).$$

The free-distance asymptote is plotted in Figure 8.7 along with simulation results using the iterative decoding algorithm of [1] with 18 iterations. This figure clearly shows that the free-distance asymptote accurately estimates the performance of the code for moderate and high SNRs.

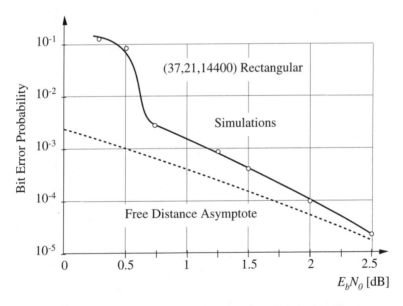

Figure 8.7 Comparison of the simulated performance with the free-distance asymptote for a Turbo code with a rectangular interleaver.

This code achieves a bit error rate of 10^{-5} at an SNR of 2.7 dB and thus performs 2 dB worse than the $(37, 21, 65536)$ Turbo code with a pseudorandom interleaver even though it has a much larger free distance. The relatively poor performance of the $(37, 21, 14400)$ Turbo code with a rectangular interleaver is due to the large multiplicity of d_{free} paths. This results in an effective multiplicity of

$$\frac{N_{\text{free}}}{N} = \frac{28,900}{14,400} \approx 2,$$

which is much larger than the effective multiplicity of the $(37, 21, 65536)$ Turbo code. We now show that the large multiplicity is a direct consequence of the use of the rectangular interleaver and that, furthermore, increasing the size of the interleaver does not result in a significant reduction in the effective multiplicity of the free-distance codewords.

The free-distance paths in the Turbo code with the rectangular interleaver are due to four basic information sequences of weight 4. These information sequences are depicted in Figure 8.8 as they would appear in the rectangular interleaver. The "square" sequence in Figure 8.8a depicts the sequence $\mathbf{u} = 1, 0, 0, 0, 0, 1, 0_{594}, 1, 0, 0, 0, 0, 1, 0_{\infty}$, where 0_{594} denotes a sequence of 594 consecutive zeros and 0_{∞} represents a sequence of zeros that continues to the end of the information sequence. In this case, the rectangular interleaver maps the sequence $\mathbf{u}$ to itself and therefore $\mathbf{u}' = \mathbf{u}$. The sequence $\mathbf{u}$ results in a parity sequence $\mathbf{v}^1$ from the first constituent encoder that, after puncturing, has weight 4. Similarly, the input sequence $\mathbf{u}' = \mathbf{u}$ results in a parity sequence $\mathbf{v}^2$ from the second constituent encoder that, after puncturing, also has weight 4. The weight of the codeword is then $d_{\text{free}} = 4 + 4 + 4 = 12$. Since the "square" sequence in Figure 8.8a can appear in $(\sqrt{N} - 5) \times (\sqrt{N} - 5) = 13,225$ distinct positions in the rectangular interleaver, and in each case $\mathbf{u}' = \mathbf{u}$ and a codeword of weight $d_{\text{free}} = 12$ results, this results in 13,325 free-distance codewords. Note that for every occurrence of the "square" sequence to result in a codeword of weight 12 the weight of both parity sequences must be invariant to which is punctured first.

The "rectangular" sequences in Figures 8.8b and 8.8c also result in weight-12 codewords. For these two sequences, the weight of one of the parity sequences is affected by whether or not it is punctured first and only every other position in which the "rectangular" sequences appear in the interleaver results in a codeword of weight $d_{\text{free}} = 12$. Thus, the sequences in Figure 8.8b and Figure 8.8c each result in $0.5(\sqrt{N} - 10) \times (\sqrt{N} - 5) = 6325$ free-distance codewords. For the sequence in Figure 8.8d, the weight of both parity sequences is affected by which is punctured first, and only one out of four positions in which the "rectangular" sequence appears in the interleaver results in a codeword of weight $d_{\text{free}} = 12$. Consequently, this sequence results in $0.25(\sqrt{N} - 10) \times (\sqrt{N} - 10) = 3025$ free-distance

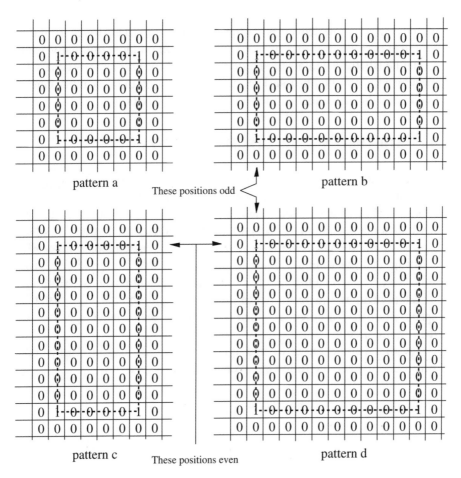

Figure 8.8 Information sequences causing d_{free} codewords in a Turbo code with a rectangular interleaver.

codewords. Summing the contributions of each type of sequence results in a total of $N_{\text{free}} = 28,900$ codewords of weight $d_{\text{free}} = 12$.

It is tempting to try to improve the performance of a Turbo code with a rectangular interleaver by increasing the size of the interleaver. However, all of the information sequences shown in Figure 8.8 would still occur in a larger rectangular interleaver, so the free distance cannot be increased by increasing N. Also, since the number of free-distance codewords is on the order of N, increasing the size of the interleaver results in a corresponding increase in N_{free} such that the effective multiplicity N_{free}/N does not change significantly. Without the benefit of a reduced effective multiplicity, the free-distance asymptote, and thus the error floor, of Turbo codes with rectangular interleavers is not lowered enough for them to manifest the excellent performance of Turbo codes with pseudorandom interleavers for moderate BERs. Attempts to design interleavers for Turbo codes generally introduce structure to the interleaver and thus destroy the very randomness that results in such excellent performance at low SNRs.

8.6   THE DISTANCE SPECTRUM OF TURBO CODES

In the previous section it was shown that the error floor observed in the performance of Turbo codes is due to their relatively low free distance. It is now shown that the outstanding performance of Turbo codes at low SNRs is a manifestation of the sparse distance spectrum that results when a pseudorandom interleaver is used in a parallel concatenation scheme. To illustrate this the weight-2 distance spectrum of a $(37, 21, 65536)$ Turbo code averaged over all possible pseudorandom interleaver is found and its relationship to the performance of the code is discussed. The distance spectrum of the Turbo code is then compared to the distance spectrum of the $(2, 1, 14)$ code.

With the algorithm described in [26], the $(37, 21, 65536)$ Turbo code was found to have the following average distance spectrum:

d	N_d	W_d
6	4.5	9
8	11	22
10	20.5	41
12	75	150

where, for reasons discussed in the next section, only weight-2 information sequences are considered. The distance spectrum information for a distance d is referred to as a spectral line. This data can be used in conjunction with the bound of equation (8.10) to estimate the performance of the code. (Note that $W_d = N_d \tilde{w}_d$.) In addition, by plotting each term of equation (8.10) the contribution of each spectral line to the overall performance of the code can be estimated.

The bound on the performance of the Turbo code with the preceding distance spectrum is given in Figure 8.9 along with curves showing the contribution of each spectral line. This clearly shows that the contribution to the code's BER by the higher-distance spectral lines is less than the contribution of the free-distance term for SNRs greater than 0.75 dB. Thus, the free-distance asymptote dominates the performance of the code not

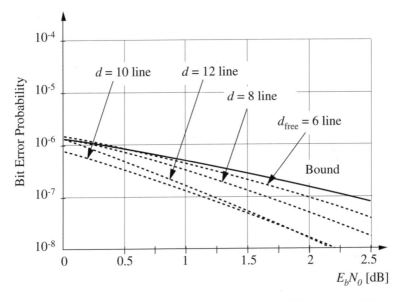

Figure 8.9 Performance bounds and the influence of the different spectral lines for an average (37, 21, 65536) Turbo code.

only for moderate and high SNRs, but also for low SNRs. We characterize distance spectra for which this is true as sparse or spectrally thin.

The ramifications of a sparse distance spectrum are made evident by examining the distance spectrum of convolutional codes. The (2, 1, 14) convolutional code introduced in Section 8.1, has the following distance spectrum:

d	N_d^0	W_d^0
18	33	187
20	136	1034
22	835	7857
24	4787	53994
26	27941	361762
28	162513	2374453
30	945570	15452996
32	5523544	99659236

as reported in [29]. When comparing the distance spectrum of a convolutional code and a Turbo code, it is important to remember that for a convolutional code $N_d \approx N \times N_d^0$ for the low-weight codewords.

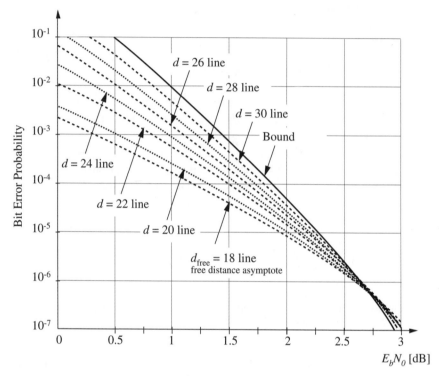

Figure 8.10 Performance bounds and the influence of the different spectral lines for a (2, 1, 14) convolutional code.

Figure 8.10 shows the estimated performance of this code using the bound of equation (8.9) and the contribution of each spectral line.

In this case the contribution of the higher-distance spectral lines to the overall BER is greater than the contribution of the free-distance term for SNRs less than 2.7 dB, which corresponds to BERs of less than 10^{-6}! The large SNR required for the free-distance asymptote to dominate the performance of the (2, 1, 14) code is due to the rapid increase in the path multiplicity for increasing d. We characterize distance spectra for which this is true as spectrally dense. The dense distance spectrum of convolutional codes also accounts for the discrepancy between the real coding gain at a particular SNR and the asymptotic coding gain calculated using just the free distance [28].

Thus, it can be concluded that the outstanding performance of Turbo codes at low signal-to-noise ratios is a result of the dominance of the free-distance asymptote, which in turn is a consequence of the sparse distance spectrum of Turbo codes, as opposed to spectrally dense convolutional codes. Finally, the sparse distance spectrum of Turbo codes is due to the structure of the codewords in a parallel concatenation and the use of pseudorandom interleaving.

8.7   SPECTRAL THINNING

In this section the observations made concerning the distance spectrum and spectral thinning of Turbo codes are formalized from the point of view of random interleaving.[1] Random interleaving was introduced in [5] and [6] to develop transfer function bounds on the average performance of Turbo codes. Here, random interleaving is used to explore the effect of the interleaver on the distance spectrum of the code. To simplify the notation and discussion, only nonpunctured Turbo codes are considered explicitly. The extension to punctured codes is straightforward and may be found in [26].

The fundamental idea of random interleaving is to consider the performance of a Turbo code averaged over all possible pseudorandom interleavers of a given length. For a given N there are $N!$ possible pseudorandom interleavers and, assuming a uniform distribution, each occurs with probability $1/N!$. Let a particular interleaver map an information sequence $\mathbf{u}$ of weight w to an information sequence $\mathbf{u}'$, also of weight w. Then there are a total of $w!(N-w)!$ interleavers in the ensemble of $N!$ interleavers that perform this same mapping. Thus, the probability that such a mapping— and, hence, that the codeword that results from the input sequences $\mathbf{u}$ and $\mathbf{u}'$—occurs is

$$\frac{w!(N-w)!}{N!} = \frac{1}{\binom{N}{w}}.$$

Following [7], define the input redundancy weight enumerating function (IRWEF) of a systematic code as

$$A(W, Z) = \sum_{w}\sum_{z} A_{w,z} W^w Z^z, \tag{8.13}$$

where $A_{w,z}$ is the number of codewords of weight $d = w + z$ generated by input sequences of weight w and parity sequences of weight z. The goal

---

[1] In the rapidly developing field of Turbo codes, the development presented in this section is but one point of view. The reader is encouraged to peruse the literature for alternative analyses. In particular, the work of Benedetto and Montorsi [6, 8] and Divsilar, et al. [14–17] has significantly advanced the understanding of Turbo codes.

is now to develop a relationship between the codewords in the constituent encoders and $A_{w,z}$ for the Turbo code and to see how that relationship changes with the size of the interleaver.

Recall from Section 8.2 that a Turbo codeword is essentially the combination of a codeword from the first constituent encoder plus a codeword from the second constituent encoder. A codeword of weight $d_1 = w + z_1$ from the first constituent encoder caused by an information sequence **u** of weight w is composed of n_1 detours of total length l_1. The ordered set of n_1 detours in the first encoder is denoted by S_1. The information sequence **u** that results in the set S_1 is mapped by a particular interleaver to the information sequence **u′**, also of weight w, which is then encoded by the second constituent encoder. This results in a codeword of weight $d_2 = w + z_2$, with the ordered set S_2 consisting of n_2 detours of total length l_2.

For example, Figure 8.3 depicts a codeword of weight $d_1 = 4 + 3 = 7$ in the first constituent encoder caused by a information sequence of weight $w = 4$ and composed of $n_1 = 2$ detours of total length $l_1 = 5 + 3 = 8$. For the interleaver described in Section 2, **u** is mapped to an **u′** that results in a codeword of weight $d_2 = 4 + 3 = 7$ in the second constituent encoder consisting of $n_2 = 2$ detours of total length $l_2 = 3 + 3 = 6$. Thus, this S_1 and S_2 result in a codeword of weight 10 and a single contribution to $A_{4,6}$ of the Turbo code for this particular interleaver. Averaged over the ensemble of interleavers of length N, a set S_1 with information weight w and parity weight z_1 and a set S_2 with information weight w and parity weight z_2 will contribute a fraction

$$\frac{1}{\binom{N}{w}}, \tag{8.14}$$

to the enumerating function coefficients A_{w,z_1+z_2} of "average" Turbo code.

Because the sequence of zeros connecting any two distinct detours has no effect on the weight of the information sequence or the parity sequence, there are

$$\binom{N - l_1 + n_1}{n_1} \tag{8.15}$$

ways that the ordered set, S_1, of n_1 detours can be arranged such that their contribution to A_{w,z_1+z_2} is not changed. This is simply the number of ways in which n_1 distinct detours can be arranged in a sequence of length N while maintaining the order in which they appear. Similarly, if the codeword in the second constituent encoder ends in the all-zero state, then the ordered set S_2 will make

$$\begin{pmatrix} N - l_2 + n_2 \\ n_2 \end{pmatrix} \tag{8.16}$$

contributions to A_{w,z_1+z_2}. However, because the second encoder is not guaranteed to return to the all-zero state, it is possible that the last of the n_2 detours is not actually a detour, but instead ends in a nonzero state. In this case, the last detour cannot be moved and the set S_2 makes

$$\begin{pmatrix} N - l_2 + (n_2 - 1) \\ (n_2 - 1) \end{pmatrix} \tag{8.17}$$

contributions to A_{w,z_1+z_2}.

The contribution to the distance spectrum of a Turbo code due to any pair of ordered sets S_1 and S_2, averaged over the ensemble of pseudorandom interleavers, can now be computed using equations (8.14), (8.15), (8.16), and (8.17). If the codeword in the second constituent encoder happens to end in the all-zero state, then the contribution of some S_1 and S_2 to A_{w,z_1+z_2} is given by

$$\frac{\begin{pmatrix} N - l_1 + n_1 \\ n_1 \end{pmatrix} \begin{pmatrix} N - l_2 + n_2 \\ n_2 \end{pmatrix}}{\begin{pmatrix} N \\ w \end{pmatrix}}. \tag{8.18}$$

If the codeword in the second constituent encoder does not end in the all-zero state, then the contribution to A_{w,z_1+z_2} is given by

$$\frac{\begin{pmatrix} N - l_1 + n_1 \\ n_1 \end{pmatrix} \begin{pmatrix} N - l_2 + (n_2 - 1) \\ (n_2 - 1) \end{pmatrix}}{\begin{pmatrix} N \\ w \end{pmatrix}}. \tag{8.19}$$

Equations (8.18) and (8.19) can now be used to explore the effect of changing the interleaver size on the distance spectrum of an "average" Turbo code.

Since we are primarily concerned with low-weight codewords in the distance spectrum, we assume that $N \gg n_1, n_2, l_1,$ and l_2. If this is not true, then S_1 and S_2 either contain a very large number of short detours or a few very long detours. In both cases, it is very unlikely that the result is a codeword of low weight. With this assumption, equation (8.18) can be approximated by

$$\frac{w!}{n_1! n_2!} N^{n_1+n_2-w}, \tag{8.20}$$

where, without loss of generality, it is assumed that $n_1 \geq n_2$. Since each detour is caused by an information sequence of weight at least 2, $w \geq 2n_1$.

The behavior of equation (8.20) for increasing N can be broken down into three cases:

1 $n_1 > n_2$: The exponent of N is strictly negative and the contribution to A_{w,z_1+z_2} decreases as N increases.

2 $n_1 = n_2$ and $w > 2n_1$: The exponent of N is strictly negative and the contribution to A_{w,z_1+z_2} decreases as N increases.

3 $n_1 = n_2$ and $w = 2n_1$: The exponent of N is exactly zero and the contribution to A_{w,z_1+z_2} converges to a finite value as N increases.

Similarly, for $N \gg n_1, n_2, l_1$, and l_2, equation (8.19) can be approximated by

$$\frac{w!}{n_1!(n_2-1)!} N^{n_1+n_2-w-1}, \tag{8.21}$$

where $w \geq 2n_1$. However, since the tail in the second encoder may be caused by an information sequence of weight 1, we also have $w \geq 2n_2 - 1$. The behavior of equation (8.21) for increasing N can be broken down into three cases:

1 $n_1 > n_2$: Since $w \geq 2n_1$, the exponent of N is strictly negative and the contribution to A_{w,z_1+z_2} decreases as N increases.

2 $n_1 = n_2$: Again, since $w \geq 2n_1$, the exponent of N is strictly negative and the contribution to A_{w,z_1+z_2} decreases as N increases.

3 $n_1 < n_2$: Since $w \geq 2n_2 - 1$, the exponent of N is strictly negative and the contribution to A_{w,z_1+z_2} decreases as N increases.

The following lemma can now be stated.

LEMMA 8.1

Given a Turbo code based on two systematic feedback encoders and a pseudorandom interleaver of length N in which the first encoder is assumed to be forced back to the all-zero state, the contribution of two ordered sets of detours S_1 and S_2 with the same information weight to the distance spectrum of the Turbo code, averaged over all pseudorandom interleavers of length N, converges to a nonzero constant as $N \to \infty$, if and only if

1 S_2 leaves the second encoder in the all-zero state.

2 S_1 and S_2 contain the same number of detours.

3 Each detour in S_1 and S_2 is caused by a weight-2 information sequence.

In all other cases, the contribution goes to zero as $N \to \infty$.

For each term $A_{w,z}$ in the IRWEF of a Turbo code there is a finite number of pairs of sets S_1 and S_2 that contribute to it. If either set contains a long detour, then it is possible that that pair will be excluded for small interleavers. As the size of the interleaver increases, eventually all pairs of sets will be allowed and any further increase in N will not result in additional pairs of sets contributing to $A_{w,z}$. Also, as $N \to \infty$, $A_{w,z}$ will be dominated by pairs of S_1 and S_2 that satisfy the three conditions of Lemma 8.1, and thus each $A_{w,z}$ will converge to a finite value. Since each spectral line is a finite sum of $A_{w,z}$ terms, each spectral line converges to a finite value as the interleaver size increases.

The convergence of each spectral line to a finite value as the size of the interleaver increases results in spectral thinning. That is, for small interleavers there may be pairs of sets S_1 and S_2 that do not satisfy the conditions of Lemma 8.1, but which contribute to the multiplicity of a low-weight spectral line. As the size of the interleaver increases, the number of these aberrational sets decreases until the spectral line reaches its final value as determined by Lemma 8.1. It is this thinning of the distance spectrum that enables the free-distance asymptote of a Turbo code to dominate the performance for low SNR and thus to achieve near-capacity performance.

8.8   PRIMITIVE POLYNOMIALS AND FREE DISTANCE

We now consider the ramifications of Lemma 8.1 with respect to the free-distance codewords of a Turbo code. That is, what does Lemma 8.1 imply about the information sequences that generate the free-distance codewords in an "average" Turbo code?

For an average Turbo code, Lemma 8.1 states that as the size of the interleaver increases each spectral line is the result of contributions only from pairs of ordered sets of detours in which each detour is caused by a weight-2 information sequence. It is reasonable to expect that the free-distance spectral line will be among the first spectral lines to converge to its final value. Thus, for reasonably large interleavers the free distance will be determined by the sets S_1 and S_2 satisfying the conditions of Lemma 8.1. Let s_1 and s_2 be any pair of detours caused by a weight-2 information sequence that results in a minimum-weight parity sequence in the first and second constituent encoders, respectively. Note that there may be more than one such pair of minimum-weight detours for the constituent encoders.

A free-distance codeword in an average Turbo code must be the result of sets S_1 and S_2 that consist of only those minimum-weight detours

s_1 and s_2, respectively. Furthermore, since each additional detour in either S_1 and S_2 adds weight to the codeword, S_1 and S_2 must each contain only one minimum-weight detour. (If each detour does not add weight, then a weight-2 information sequence exists that generates a zero-weight parity sequence and the free distance of the code would be 2.) Thus, the free-distance codewords of an average Turbo code are caused by weight-2 information sequences, provided the interleaver is large enough.

Therefore we have the following.

LEMMA 8.2

For an average Turbo code, as the size of the interleaver approaches infinity:

1 The free-distance codewords are caused by information sequences of weight 2.

2 The free distance of an average Turbo code is maximized by choosing constituent encoders that have the largest output weight for weight-2 information sequences.

We now present an intuitive argument for why choosing the feedback polynomial $h_0(D)$ in a $(2, 1, v)$ systematic feedback encoder to be a primitive polynomial maximizes the output weight for weight-2 information sequences. It follows that the free distance of an average Turbo code is maximized by using a primitive polynomial as the feedback polynomial in the constituent encoders.

The generator matrix of a $(2, 1, v)$ systematic feedback encoder is

$$G(D) = \left[1 \quad \frac{h_1(D)}{h_0(D)} \right],$$

where $h_1(D)$ and $h_0(D)$ are referred to as the feedforward and feedback polynomials, respectively, and $h_0(D)$ is of degree v. Since only information sequences of weight 2 are being considered, the systematic output contributes weight 2 to the overall codeword weight for all the encoders being considered. Therefore, only the weight contributed by the parity sequence, that is,

$$v(D) = u(D) \frac{h_1(D)}{h_0(D)},$$

needs to be maximized. Furthermore, since we are concerned only with the choice of $h_0(D)$, $h_1(D)$ is assumed to be a polynomial such that $h_0(D)$ and $h_1(D)$ are relatively prime. (There is empirical evidence that the choice of both polynomials can affect the performance of the code [26, 27], but we will not address that issue here.)

Let $1 + D^K$, for some finite K, be the shortest input sequence of weight 2 that generates a finite-length codeword. The resultant parity sequence is

$$
\begin{aligned}
v(D) &= (1 + D^K)\frac{h_1(D)}{h_0(D)} \\
&= \frac{h_1(D)}{h_0(D)} + D^K\frac{h_1(D)}{h_0(D)}.
\end{aligned}
$$

Since $v(D)$ is of finite length K, $h_1(D)/h_0(D)$ must be periodic with period K. Increasing the period K increases the length of the shortest weight-2 input sequence that generates the finite-length sequence $v(D)$ and therefore increases the length of $y(D)$. Intuitively, one would expect that increasing its length would result in $y(D)$ gaining weight. That is, on average, half of the added bits would be ones.

A strictly proper rational function of two polynomials, such as $h_1(D)/h_0(D)$, is periodic with period $K \le 2^v - 1$. The period is maximized, that is, $K = 2^v - 1$, when $h_0(D)$ is a primitive polynomial. Since the free distance of the average Turbo code is determined by information sequences of weight 2, for sufficiently large interleavers the free distance will be maximized by maximizing K. Therefore, choosing $h_0(D)$ to be a primitive polynomial will result in a larger free distance for an average Turbo code.

To test this, we compare a $(37, 21, 400)$ Turbo code which has $K = 5$ and free distance $d_{\text{free}} = 6$ to a $(23, 35, 400)$ Turbo code. Both codes are punctured as in [1]. The feedback polynomial $h_0 = 23$ in the second Turbo code is a primitive polynomial of degree $v = 4$, and thus $1/h_0(D)$ has a period of $K = 2^v - 1 = 15$. The free distance of this Turbo code was found to be $d_{\text{free}} = 10$ [9, 26]. Figure 8.6 shows simulation results for these two codes using the iterative decoding algorithm of [1] with 18 iterations. As expected, the second Turbo code performs better at moderate and high SNRs because its free-distance asymptote is steeper due to the increased free distance.

8.9   CONCLUSION

The exceptional performance of Turbo codes at low SNRs is due to the sparse distance spectrum and the resultant ability of the code to follow the free-distance asymptote at moderate to low SNRs. The use of systematic feedback encoders and pseudorandom interleavers results in spectral thinning, in which information sequences that generate low-weight parity sequences from the first constituent encoder are interleaved with high probability to information sequences that generate high-weight parity sequences in the second constituent encoder. Spectral thinning is enhanced by increasing interleaver lengths. For very large interleavers, spectral thinning results in a sparse distance spectrum in which the first several spectral lines are determined solely by input sequences of weight 2. Thus, spectral thinning results in few low-weight codewords and a large number of codewords of "average" weight. This is very similar to the type of distance spectrum achieved by "random-like" codes [4].

In a more philosophical light, Turbo codes remind us that information-theoretical arguments imply that long block lengths, but not necessarily large free distances, are required to achieve capacity at moderate BERs. Thus, like convolutional codes, Turbo codes are a class of codes that achieve long block lengths, but without the corresponding increased density of the distance spectrum common to convolutional codes, and for which a practical, albeit nontrivial, decoding algorithm exists. In addition, Turbo codes are time varying due to the pseudorandom interleaver, and the time-varying structure is essential in achieving the distance spectrum that results in near-capacity performance at moderate BERs. This suggests that some effort should be made to find other classes of time-varying codes, and decoding algorithms, that have good distance spectra rather than just large free distances.

REFERENCES

[1] C. Berrou, A. Glavieux, and P. Thitimajshima, "Near Shannon limit error-correcting coding and decoding: Turbo-codes," *Proc. 1993 IEEE Int. Conf. on Comm.*, Geneva, Switzerland, pp. 1064–1070, 1993.

[2] C. Berrou and A. Glavieux, "Turbo-codes: general principles and applications," *Proc. 6th Tirrenia Int. Workshop on Digital Communications*, Tirrenia, Italy, pp. 215–226, September 1993.

[3] G. Battail, C. Berrou, and A. Glavieux, "Pseudo-random recursive convolutional coding for near-capacity performance," *Proc. Comm. Theory Mini-Conf.*, Globecom '93, Houston, pp. 23–27, December 1993.

[4] G. Battail, "On random-like codes," unpublished manuscript.

[5] S. Benedetto and G. Montorsi, "Average performance of parallel concatenated block codes," *Electron. Lett.*, Vol. 31, No. 3, pp. 156–158, 1995.

[6] S. Benedetto and G. Montorsi, "Performance evaluation of TURBO-codes," *Electron. Lett.*, Vol. 31, No. 3, pp. 163–165, 1995.

[7] S. Benedetto and G. Montorsi, "Unveiling turbo codes: some results on parallel concatenated coding schemes," *IEEE Trans. Inform. Theory*, Vol. IT-42, No. 2, pp. 409–428, 1996.

[8] S. Benedetto and G. Montorsi, "Design of parallel concatenated convolutional codes," *IEEE Trans. Commun.*, Vol. COM-44, No. 5, pp. 591–600, 1996.

[9] P. Robertson, "Illuminating the structure of parallel concatenated recursive systematic (TURBO) codes," *Proc. GLOBECOM '94*, Vol. 3, pp. 1298–1303, San Francisco, November 1994.

[10] J. Hagenauer, E. Offer, and L. Papke, "Iterative decoding of binary block and convolutional codes," *IEEE Trans. Inform. Theory*, Vol. IT-42, No. 2, pp. 429–445, 1996.

[11] A. S. Barbulescu and S. S. Pietrobon, "Interleaver design for turbo codes," *Electron. Lett.*, Vol. 30, No. 25, p. 2107, 1994.

[12] A. S. Barbulescu and S. S. Pietrobon, "Terminating the trellis of turbo-codes in the same state," *Electron. Lett.*, Vol. 31, No. 1, pp. 22–23, 1995.

[13] A. S. Barbulescu and S. S. Pietrobon, "Rate compatible turbo codes," *Electron. Lett.*, Vol. 31, No. 7, pp. 535–536, 1995.

[14] D. Divsalar and F. Pollara, "Turbo codes for PCS applications," *Proc. 1995 IEEE Int. Conf. on Communications*, Seattle, WA, June, 1995.

[15] D. Divsalar and F. Pollara, "Turbo codes for deep-space communications," *JPL TDA Progress Report 42-120*, February 1995.

[16] D. Divsalar, S. Dolinar, F. Pollara, and R. J. McEliece, "Transfer function bounds on the performance of turbo codes," *JPL TDA Progress Report 42-122*, pp. 44–55, August 1995.

[17] S. Dolinar and D. Divsalar, "Weight distributions for Turbo codes using random and nonrandom permutations," *JPL TDA Progress Report 42-122*, pp. 56–65, August 1995.

[18] R. J. McEliece, E. R. Rodemich, and J.-F. Cheng, "The Turbo decision algorithm," *Proc. 33rd Ann. Allerton Conf. on Communication, Control and Computing*, Monticello, IL, October 1995.

[19] Y. V. Svirid, "Weight distributions and bounds for Turbo codes," *Eur. Trans. Telecommun.*, Vol. 6, No. 5, pp. 543–556, 1995.

[20] O. Joerssen and H. Meyr, "Terminating the trellis of turbo-codes," *Electron. Lett.*, Vol. 30, No. 16, pp. 1285–1286, 1994.

[21] P. Jung and M. Naßhan, "Performance evaluation of turbo codes for short frame transmission systems," *Electron. Lett.*, Vol. 30, No. 2, pp. 111–113, 1994.

[22] P. Jung and M. Naßhan, "Dependence of the error performance of turbo-codes on the interleaver structure in short frame transmission systems," *Electron. Lett.*, Vol. 30, No. 4, pp. 285–288, 1994.

[23] P. Jung, "Novel low complexity decoder for turbo-codes," *Electron. Lett.*, Vol. 31, No. 2, pp. 86–87, 1995.

[24] G. Caire, G. Taricco, and E. Biglieri, "On the convergence of the iterated decoding algorithm," *Proc. 1995 IEEE Int. Symp. on Information Theory*, Whistler, BC, Canada, p. 472, September 1995.

[25] N. Wiberg, H.-A. Loeliger, and R. Kötter, "Codes and iterative decoding on general graphs," *Eur. Trans. Telecommun.*, Vol. 6, pp. 513–525, 1995.

[26] J. Seghers, *On the Free Distance of TURBO Codes and Related Product Codes*, Final Report, Diploma Project SS 1995, Number 6613, Swiss Federal Institute of Technology, Zurich, Switzerland, August 1995.

[27] D. Arnold and G. Meyerhans, *The Realization of the the Turbo-Coding System*, Swiss Federal Institute of Technology, Zurich, Switzerland, 1995.

[28] S. Lin and D. J. Costello, Jr., *Error Control Coding: Fundamentals and Applications*, Prentice-Hall, NJ, 1983.

[29] M. Cedervall and R. Johannesson, "A fast algorithm for computing distance spectrum of convolutional codes," *IEEE Trans. Inform. Theory*, IT-35, pp. 1146–1159, 1989.

INDEX

C

ABOUT THE AUTHOR

Christian B. Schlegel was born and raised in St. Gallen, Switzerland. After completing the Dipl. El. Ing. ETH degree at the Federal Institute of Technology (ETH) in Zürich in 1984, and gaining some experience in local industry, he travelled to the United States and enrolled at the University of Notre Dame, Indiana in 1985. There he obtained the M.S. and Ph.D. degrees in electrical engineering in 1986 and 1988, respectively, and was first exposed to trellis coding under the supervision of his advisor, Professor Daniel J. Costello, Jr.

In 1988 he joined the Communications Group at the research center of Asea Brown Boveri, Ltd., in Baden, Switzerland, working mainly on mobile communications research projects. During the 1991–1992 academic year, he visited the University of Hawaii at Manoa, Hawaii, after which he joined the Digital Communications Group at the University of South Australia in Adelaide as head of the Mobile Communications Research Centre. During that time his work expanded to include multiple-access communications.

In 1994 he joined the Division of Engineering at the University of Texas in San Antonio, where he started the Communications, Coding, and Control Laboratory, continuing work on problems in multiple-access communications, trellis coding, and mobile radio systems. His work is widely published and he consults for and cooperates with diverse national and international industrial and academic partners. His research is supported by various research grants. *Trellis Coding* is Dr. Schlegel's first completed research monograph, and he is currently working with Dr. Alex Grant on a book titled *Coordinated Multiple User Communications* in production at Kluwer Academic Publisher.

Dr. Schlegel's varied interests include skiing and rock climbing, as well as analytical psychology, mythology, language, and poetry. In 1996, the National Library of Poetry was the first to publish one of his poems.